The Complete
Offshore Yacht

Published by
YACHTING MONTHLY
IPC Magazines Ltd, King's Reach Tower, Stamford Street,
London SE1 9LS. 1990

Copyright © IPC Magazines Ltd, 1990

ISBN 1 – 85277 – 068 – 6

All rights reserved. No part of this publication may be reproduced, stored in a retrieval system,
or transmitted, in any form or by any means, electronic, mechanical, photocopying, recording or
otherwise, without the prior permission of the publisher

Neither the publisher nor the authors can accept responsibility for errors,
omissions or alterations in this book.

The Complete Offshore Yacht is printed by Pensord Press Ltd, Tram Road, Pontllanfraith, Blackwood. Gwent.
Colour reproduction by Marlin Graphics Ltd, Sidcup, Kent

The Complete
Offshore Yacht

YACHTING MONTHLY

Contributing authors

Bill Anderson, Andrew Bray, Tony Castro, Kitty Hampton, Peter Haward, James Jermain,
Ian Nicolson, Geoff Pack, Chuck Paine, David Thomas,
The Wolfson Unit, Southampton University

Illustrators

Mike Collins, Richard Hawke, John Moxham,
Ian Nicolson, Arthur Saluz

Photographs

As credited on individual pictures

Design

Caroline Helfer

Cover

YM Offshore 35, illustrated by Arthur Saluz
Back cover: YM Offshore 35, cutaway illustration by John Moxham

Contents

Chapter One
Fit for offshore ?   1
Definitions of seaworthiness and the offshore yacht

Chapter Two
The designers speak   10
Tony Castro, Chuck Paine and David Thomas each outline the thinking behind their offshore yachts

Chapter Three
Watertight integrity   18
A close look at every opening in the structural monocoque

Chapter Four
The seaworthy rig   24
Rigs and rigging have changed enormously in recent years, and here we cull through good and bad to create the most practical and efficient

Chapter Five
Decks – the working platform   30
Working the offshore yacht efficiently requires careful initial design and subsequent layout and equipping

Chapter Six
Into the interior   36
The accommodation of the offshore yacht has to work efficiently in two arenas, in harbour and at sea. The harbour part is easy, here we look at features that work at sea.

Chapter Seven
Engines and engine reliability   42
The modern diesel is central to the running of the offshore yacht. Various features will improve the chances of it running better and longer

Chapter Eight
Fastnet – Ten years on   48
    The Storm – a new theory   49
    Storm damage   51
    The stability factor   53
    Windward ability – drive vs drag   58
    Clipped on – but are you safe ?   60
    Liferafts – the lessons learned   63
    Survey results   66

Chapter Nine
Electrical systems   70
Nowhere has there been more change in recent years than in the electrical systems of yachts

Chapter Ten
Equipping for emergencies   76
The priorities for the safe running of a yacht sailing offshore are examined in detail

Chapter Eleven
The Yachting Monhtly Offshore 35   81
Distilling as much of the information of the previous ten chapters as possible, a design is commissioned to embody the best features

Introduction

The disastrous Fastnet Race of 1979 marked a cornerstone in yacht design and equipping. In the following ten years yachts changed a good deal, but was it for the better or worse? Have salesmen more influence than seamen in the design equation, has commercial competition diminished the seaworthiness of the end product or has new technology in terms of design and materials created inherently better boats?

During the course of 1989 Yachting Monthly magazine gathered together a team of some of the most highly respected professionals in yachting, each an expert in their field, to assemble and publish what is the most thorough and detailed appraisal of the modern offshore yacht ever compiled. Every element of the offshore yacht was examined in great detail from the complicated compromise of design considerations down to small details such as the correct way of attaching harness U-bolts (destruction tests were carried out on eight methods to ascertain their effectiveness).

The magazine series was widely praised and now The Complete Offshore Yacht compiles the whole year-long series into book format.

The Offshore Yacht Advisory Panel

Bill Anderson
Training Manager of the RYA and a regular contributor to *Yachting Monthly*, Bill is a highly experienced yachtsman and devises the RYA Yachtmaster scheme. He was a key figure in the Fastnet Race Report.

Peter Haward
Peter is a professional yacht delivery skipper and must rank as one of the most experienced deep sea yachtsmen afloat today. His experience of hundreds of different craft in widely varying conditions from inshore gales to offshore hurricanes is invaluable to the book.

Ian Nicolson
Naval architect and surveyor as well as accomplished author. He has also notched up an impressive number of long-distance ocean cruises, was a founder member of the Ocean Cruising Club, and has a strong practical streak running through his approach to sailing and writing.

David Thomas
A highly successful naval architect with dozens of well-known designs to his name, ranging from boats like the Elizabethan 31 to the Hunter Sonata, Sigma 33 and Horizon 32. David was Assistant Editor of *Yachting World* for many years and completed the 1979 Fastnet Race in a Sigma 33.

The Wolfson Unit
The Wolfson Unit of Southampton University has, over a number of years, established itself as a technical authority on all aspects of small craft design through its extensive testing facilities. It is highly respected world-wide and was involved in the compilation of the 1979 Fastnet Race Report.

About the authors

Andrew Bray
Editor of Yachting Monthly and currently owns and sails a Sadler 34 out of Lymington. He has been sailing for nearly 30 years and his offshore sailing experience includes taking part in the Singlehanded Transatlantic Race, Round Britain Race, two Azores and Back Races and two Yachting Monthly Triangle Races in addition to extensive cruising in Northern European Waters, the Mediterranean and Caribbean

James Jermain
A lifetime yachtsman and yachting writer with 14 years experience. He has raced whalers in Plymouth Sound and half tonners in RORC events. He has cruised as far afield as Sweden, Tunisia and the Caribbean in addition to home waters from the Clyde to The Morbihan.

He became a yachting writer in 1974 and joined *Yachting Monthly* as Technical Editor in 1981 becoming Deputy Editor in 1985. While with YM he has sailed over 200 different types of yacht from 18 to 60ft

Geoff Pack
Assistant Editor of *Yachting Monthly* he has cruised since childhood in a wide variety of craft from Thames barges to high tech multihulls in many parts of the world.

Joined YM after college in 1977 but had a four year break to go ocean cruising, during which time he sailed approximately 40,000 miles across (three times) and around the Atlantic. Currently owns **Foreigner**, a 41ft Sailcraft Apache catamaran which he sails with his young family.

Chapter 1

Fit for offshore?

One boat's hard sail might be another's survival conditions. This chapter examines definitions of seaworthiness

A REASONED argument states that a well corked bottle will survive *any* storm and it is a sound philosophy when it comes to the building and fitting out of yachts. However, drop that bottle in the middle of the North Sea or English Channel in those conditions and very quickly it will be smashed to pieces on the nearest beach downwind. The ability to stay afloat is, of course, an essential element when one talks of seaworthiness and in this one deals with the large subject of the 'corks' – every opening in the structural monocoque. However, as important is the ability of the boat through her design, construction, layout and equipment to allow her crew to function and therefore work her. In simple terms, able to function she will not join the ranks of broken bottles on the beach.

Just what is the Offshore Yacht?

This book is dealing with seaworthiness and strong sailing conditions. In that we do not say 'gale' because a Force 6 in a 22-footer can be the equivalent of a foaming Force 9 in a 45-footer.

On the assumption that a yacht will not set out in her gale conditions, one has to assume she will be caught out and, in that respect, any boat making a passage that is more than six hours (the 'imminent' period of the shipping forecast) from shelter is deemed to be sailing offshore. We have to assume that the decision to be out there is determined by responsible seamanship.

The offshore yacht will be completely decked-in and fully self-draining and it is reasonable to assume she has facilities to sustain a minimum passage of between 12 and 36 hours – in other words, berths for all aboard, galley and reasonable, dry stowage.

The seaworthy equation

When considering seaworthiness, it is easy to fall into the trap of looking at a boat as a whole. Ultimately, gauging a yacht's seaworthiness depends largely on her ability to function and therefore it is the combination of dozens of features and pieces of gear that makes the difference.

Seaworthiness also needs to be put into perspective. How seaworthy does a yacht need to be when she is only likely to meet conditions which will test these features perhaps 5 per cent of the time? The answer to this really depends upon how much the incorporation of seaworthy features compromises her for the remainder of the time. Alternatively, examining the problem at the root, in commercial terms, does the incorporation of certain features, gear and equipment mean that the yacht becomes unattractive from a price or accommodation point of view?

Like everything in yachting, there has to be a compromise which is balanced *and must be assessed by the yachtsman* in terms of horses for courses. One would not normally recommend a yacht like the plywood 18ft Silhouette for regular offshore work and yet (a much modified) one has circumnavigated. The Silhouette was designed, and is highly suitable for, weekend cruising and it was never part of her designer's brief to create a boat that would power through a Force 8 in her stride.

Size is misleading, however. The Silhouette is compact and her small size lends a blatantly obvious restriction on her likely use. The more tricky situation comes on a larger boat (which we'll fictionally call the Dreamer 35) which, by virtue of her length and outward appearance, looks the part even to the fairly experienced yachtsman.

Judging a Silhouette is easy, but the Dreamer 35 is an entirely different matter. She might well be a highly popular boat with sixty or seventy afloat, but she has been designed for the use which, in reality, she gets for most of the time. This boils down to day sailing for a few hours at a time in winds that rarely exceed Force 5/6, plus the 'alternative weekend cottage' rule - comfort, privacy and stowage for four or five staying aboard over short periods. In many cases, a hull shape that will scythe upwind in any weather is unlikely, in this size of yacht, to accommodate twin double aftercabins and a capacious saloon.

Heavy weather pushes the structure of a boat to her limits. But in construction terms, the large majority of modern yachts afloat today should have few worries about structure in bad weather

The Dreamer 35 is therefore a superb boat for 95 per cent of the time she is used but, when she meets a genuine Force 6-7 around a headland with only 10 miles to make her destination, things change. At this point her high windage (= full standing headroom), iron ballast (= attractive price) and wide-beamed, full sections (= good accommodation) are all adding to the situation of her running out of steam.

Perhaps a little on the tender side, she needs to be reefed well down so as not to lay on her ear, and at that point she simply doesn't have the drive to make upwind. Set a little more sail to increase drive and between the upwind gripes (caused by the generous beam and therefore distorted heeled waterline shapes) she simply smashes away making a lot of leeway while life on board becomes a misery.

What about the engine though; modern diesels are extremely reliable aren't they? Volvo, for example, quote a maximum consistent heeling angle of no more than 22 degrees if their engines are to be properly lubricated, so conditions for motor-sailing do not look encouraging, even under mainsail only, although this is probably your best hope. Under power alone, the engine that can push you along at 6.5 knots in a calm might only make 2 knots directly into a Force 7, because windage drag multiplies by the square for every knot of wind increase. Motoring is not the solution.

Down below, there is only one place to sit - on the lee bunk and somebody monopolised that berth hours ago when they felt seasick. Nobody feels up to the task of cooking or brewing tea because a couple of the locker doors burst open and the galley worktops and floor are covered in sugar.

Nobody is eating or resting (let alone sleeping) and all, including the skipper, are willing the boat into her destination as quickly as possible so normal life can resume.

This scene is not particularly exaggerated and goes to prove that, whilst twice the size of the Silhouette, the Dreamer 35, in the windward context, is no more seaworthy because she cannot operate in stronger winds.

Is it unreasonable to expect this particular type of modern yacht, which is so very efficient and purposeful for 95 per cent of the time, to be able to cope with conditions that she may meet for 4.5 of the remaining 5 per cent (assuming the other 0.5 per cent are full gale conditions)? Looked at in another way, would you take your wife and children on an early-winter trip to Scotland knowing that your car couldn't handle driving on snow, on the reasonable assumption you wouldn't meet any?

So what makes a boat seaworthy?

Seaworthiness is directly related to safety. The essence of seaworthiness is a yacht that looks after the crew who in turn look after her and therefore themselves. Furthermore, whilst looking after her crew, she must be of an overall design configuration, unlike the Dreamer 35, that actually allows her to make progress in foul conditions.

Unfortunately, primary design features that aid a yacht's heavy weather ability go directly against many of the features that make her a desirable and practical boat for the remainder of the time, so one is up against the compromise again. The narrowish, low-windage, deep-draught, lead-keeled Capable 35 will keep going in almost anything, but rafted alongside the Dreamer 35 in your favourite weekend anchorage she is going to be a poor relation. It is the balancing of this compromise, the ability to maintain both acceptable seaworthiness/safety, and space and practicality that is the crux of the matter.

Looking at dire situations, the Fastnet storm of 1979 highlighted some major design problems with the yachts involved. One of the trends to emerge from the subsequent report was that nearly all the damage and abandonments came after a yacht had suffered a B2 (beyond 90 degrees) knockdown and all loss of crew happened from yachts that had experienced B2s.

The ultimate stability of a yacht, and in particular her angle of vanishing stability (ie the angle of heel at which point she will carry on going over), was an area of major concern and the Wolfson Unit at Southampton University produced GZ curves of two specific designs to illustrate the problem. These are shown in **Figure 1**. The area within the curve below the line illustrates each boat's propensity to stay upside-down once it has gone beyond its angle of vanishing stability. The half-tonner loses its stability at the alarmingly shallow angle of 116 degrees but, more worrying, it can be relied on to stay upside-down for a good period of time (until a combination of wave and inertia heels the now upturned boat beyond 65 degrees). Not only does the Contessa possess significantly higher stability, but its angle of vanishing stability is 156 degrees and the curve demonstrates there is very little chance of it staying upside-down.

Heavy weather pushes the structure of a boat to her limits. It's fair to say that, in construction terms, the large majority of modern yachts afloat today have no particular worries about structure in bad weather. Hulls are strong and they don't receive the point loading pressures one could expect in port which might rupture them. Keels are usually well engineered using materi-

A well protected cockpit is essential in any offshore yacht. The addition of spray dodgers and pram hood can make a significant contribution to the crew's welfare
Patrick Roach

als and technology which has been transformed in the last 20 years and, despite the naysayer's fears, unsupported spade rudders very rarely give trouble in bad weather assuming they don't hit anything particularly solid (ie containers, logs or whales).

In contrast, it is the gear and equipment that one has to look at very carefully. Common problems on a cheaper boat are both the primary winches and mainsheet tackle becoming underpowered in bad conditions. Being unable to set sails properly may make up to 10 degrees difference in a boat's pointing ability. In the quest for keeping down weight, and therefore costs, safety margins on scantlings tend to be trimmed so that, instead of a chainplate or mast tang being rated at, say, more than twice the maximum theoretical load, economy dictates that the safety margin may be trimmed

The essence of seaworthiness is a yacht that looks after the crew who in turn look after her and therefore themselves

Figure 1

back to perhaps 1.5 x load.

Another major component of a yacht's seaworthiness relates to the size, number and quality of all her openings. It is a subject we shall be concentrating on heavily during the course of this book.

Modern materials and equipment have improved openings enormously. It may be regarded as looking a little amateurish, but bolting a sheet of polycarbonate over a smaller aperture as a window, if properly executed, barely diminishes the strength of the moulding around it. Although better looking, an aluminium framed window is not as strong, so area becomes all the more important a criterion. Natural lighting is a desirable feature for a yacht's interior and windows are tending to get larger and larger - an unhealthy trend unless, like many French-built boats, the exterior window area is large and through bolted, but apertures are considerably smaller.

Over the last 10 years, the stern sections of yachts have been opened up for extensive accommodation possibilities, and this has changed the size and layout of cockpit lockers. Whereas once upon a time there would be up to five opening lockers around the cockpit, each with a smallish opening, today it is more common to see a single vast locker with an equally large opening lid. Once open in bad conditions (perhaps to get out the storm jib, or lashing ropes), this lid is highly vulnerable to wind or wave plucking it off its hinges, thereby exposing an enormous and potentially crippling area.

The Fastnet storm exposed design problems with the companionway and sliding hatch areas of yachts, whose integrity is tested to the limit whilst running before bad weather and especially if the yacht experiences a knockdown. Tapered hatchways resulted in the boards needing to slip only inches before they were lost. Not only that but many configurations don't allow the hatch to be fully closed and secured without the top washboard being in. In relation to companionways, the lessons of the Fastnet Race should not be forgotten, and we shall be looking into this area in detail during the series.

A boat like the Dreamer 35, which cannot make to windward in gale or near-gale force conditions, cannot be regarded as seaworthy because sooner or later she could find herself upwind of a leeshore and unable to sail off. With genoa rolled away to nothing and heavily-reefed mainsail causing her to heel badly and therefore gripe regularly, her ability to hold her own depends entirely on running the engine flat out.

In bad conditions the most reliable engines are confronted with a variety of potential problems – overheating from the water intake being out of the water for much of the time, lack of oil pressure from excessive angles of heel, dirty fuel and clogged filters from a throughly-shaken fuel tank, maybe even water getting into the fuel tank through the air breather which in these conditions is getting a thorough wetting. Not only that, but in bad conditions ropes and sheets can easily be washed over the side, and a rope around the prop in bad weather is virtually impossible to clear. Without her engine, the Dreamer 35 will very quickly become an RNLI statistic if ever she meets the situation, perhaps in a windshift as a front passes through, of *needing* to make to weather.

The functioning of a yacht at sea in bad weather stems from a number of basic factors that have to be right. First and foremost is the motion of the boat. Few people can operate aboard a boat that is tossing around wildly, whether they are trying to plot a course, prepare food, change a headsail, or use the heads.

Motion is a highly subjective matter. Owners of heavy displacement boats vow that the steady, if exaggerated, motion of their craft is the best whilst the lighter boat seems bouncier but the extent of her motion is less. A major factor is predictability and it's no secret that boats in the middle-displacement range (displ:wl ratio 290+) upwards will have a more predictable motion.

Related to motion, a yacht's stiffness under sail is an important part of the ability of the crew to operate at sea, and therefore her seaworthiness.

After motion, an important element for her crew is the yacht's dryness; first, the amount of water she has driving over her decks and secondly (directly related to the first), is how much water comes below. A warm dry crew member is an efficient one but he is unlikely to remain so if a trip to the mast or foredeck is likely to entail a boot- or arm-full of water. Heavy weather has a way, through the pressure of water against windows and fittings, of seeking out leaks that never normally exist. Sprayhoods have revolutionised the lot of the average yachtsman, but there are still many that are simply poppered to the deck with no coaming, making them efficient in driving rain, but useless when the remainder of a green sea rolls its way aft across the coachroof.

Water below decks is an inevitability in bad weather but the degree to which it affects the safe running of the yacht is a contributory factor to her seaworthiness – how heavily does she rely on her electrics, for example?

One particularly poor trait in many modern yachts is the absence of a bilge sump. **Figure 2** demonstrates why, and a boat such as the one shown isn't able to pump her bilge unless she has three independently valved and plumbed strum boxes fitted, one amidships and one each into her heeled 'bellies'.

Although it comes lower down the priority list, the laying out of a yacht's interior bears some relation to her seaworthiness. An accommodation whose broad layout, and especially detailing, doesn't allow the crew to rest or function down below will be a further morale sapper. Features that are an asset in harbour can be a liability at sea, be it the large open-plan galley, or a voluminous saloon for comfortable and uncramped in-harbour living. The closer a heads is to the centreline of a boat, the easier it will be to use on either tack, but one tucked neatly *en suite* at the yacht's point of maximum beam will be a pain in the neck.

It is not a difficult job, through detailed modifications, to improve the interior of a yacht considerably. From a safety point of view, all heavy items must be effectively strapped in or restrained. Batteries and cookers came under specific criticism in the Fastnet Report, but so did smaller items like tins, which became lethal missiles when locker fronts burst open. Extra hand-holds and pillars can make an enormous difference to **a.** help you move around the boat but **b.** to assist in specific tasks 0 pillars especially in the latter case when they allow you to hook an arm around and still have both hands free for whatever

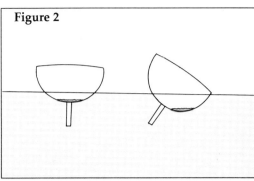

Figure 2

Patrick Roach

Although smaller yachts such as the Contessa 26 have circumnavigated their success has been due as much to the skill of their crew than to their inherent seaworthiness

you're trying to do.

The further one delves into the offshore yacht, the more one realises that it is not so much a small number of primary design considerations that make her seaworthy, but a co-ordinated union of literally hundreds of features and components.

Patrick Roach

Above left: Chuck Paine's powerful dreamship, the Bowman 40. Above: one of Tony Castro's higher profile yachts, the Barracuda 45

Chapter 2

The designers speak

Three top designers, Tony Castro, Chuck Paine and David Thomas give their views on balancing the seaworthiness equation

TONY CASTRO....

After university, Tony Castro joined Ron Holland in Ireland, where he worked for three years. He has since become one of the world's leading naval architects with a long string of racing boats and winners in his wake, including the highly successful British Admiral's Cupper *Juno*. Castro has also designed a large number of production boats from the Spring 25 and Jeanneau Arcadia to the Sun Dream and Sun Shine, up to the Sadler Barracuda.

Question: What design attributes go towards making a cruising boat suitable for offshore sailing?

The major and basic difference between sailing inshore and offshore is the remoteness from land and, obviously, the additional time it would take to get help to the yacht in distress. Problems can occur because of weather conditions and collisions with other ships, flotsam or whales.

A yacht suitable for offshore sailing is therefore one able to cope with normal sailing as well as those special conditions found offshore. Since, once caught in the middle of an offshore storm, it is impossible to avoid it, the yacht must ideally be capable of enduring it, no matter how severe.

The first priority is adequate strength of hull, deck and equipment, which nowadays should not be difficult to achieve, technically speaking.

Below: David Thomas's reputation was cemented by the evergreen and highly successful Sigma 33

Loads imposed from sea and wind conditions are calculated based on the yacht's size, displacement, stability, and rig size, and a certain amount of historical information is often used to complement the pure mathematical calculation.

The first rule, therefore, is to have the boat designed by an experienced, qualified architect.

The suitability of a yacht for offshore sailing can be subdivided into prevention and survivability. Prevention involves the decisions at the design stage of the yacht in general, the yacht's nature, its seakindliness and seaworthiness. The shape of the hull has some bearing on how it behaves in confused seas, as does its stability and the ease with which its directional control is maintained.

The most debated features are beam, displacement and stability. It is generally agreed that a wide boat with flared topsides is more likely to roll over than a narrow one with more buoyancy outboard (more slab-sided midships sections). The explanation is that the deck edge digs more easily into the water as the boat travels down the face of the wave sideways, and is more likely to trip the boat into a roll. It is also logical that a light displacement will be thrown about more violently (with greater acceleration) than a heavy displacement boat, but not necessarily with less damage. The structure in each case has to be adequate. It is by no means clear if weight by itself is more or less suitable, but rather the combination of hull shape, beam and displacement may be more relevant. Stability, everyone would agree, is always a help. The lower the centre of gravity, the less likely you are to roll over and, even if you do, provided the yacht is not unreasonably wide, it will also help to roll back up more quickly.

Survivability is partly outside the control of the designer because so many of its aspects are owner-controlled. For collisions at sea, a watertight bulkhead can sometimes help, but often the damage is not restricted to the bow area. Unsinkability is a good idea if you are prepared to pay the price. A lot of development is being done in this area and I expect improvements in one or two years.

The yacht must be prepared and maintained properly to withstand a storm and that depends entirely on the owner and skipper. The safety equipment has to be in good order and,

procedures, etc. They should also be properly clothed, with regard to the severity of the conditions. The skipper's skill and experience are most important, since we know that human error is often to blame.

Question: Have the 1979 Fastnet Race Report recommendations had any effect on the direction cruising boat design has taken since 1979?
I am sure they have had an effect, but it is difficult to quantify. One of the prime reasons why, hopefully, that tragedy will never repeat itself, is that the entry regulations on such a race have been tightened up. Some lives were lost, I am led to believe, due to premature abandonment of the yacht. Inexperience was also responsible for some loss of life.

From a design viewpoint on a world-wide basis, I'm afraid I don't think the report has had as big an impact as it should have had. Certainly, many changes have taken place, which were influenced to some extent by that event, or at least accelerated by it.

The shape of the boats currently favoured by the racing rules is not all that different from, say, the OOD 34. This boat has suffered as a result of the statistics of the disaster and yet I don't find it all that different from some of the boats being built today.

I don't think that today's production boats are built any differently except, maybe, in detail. It is true that the IOR,

> **I would always prefer an adequately strong, longer, lighter, less equipped boat than a smaller stuffed-up cruiser for the same price**

for example, hasn't changed much since then, although the introduction of the ABS guide for the construction was a well intentioned move, amongst others. A possible result of the Fastnet disaster was that, slowly, the typical production boat developed away from the IOR but, fuelled by economic pressures, it took its own direction with its own different set of problems.

At the end of the day it is difficult to impose on the public in general a boat of the shape and value-for-money they don't want. Social pressures have been responsible for the development of the production boat as we know it today to a very large extent.

Financial pressures, still higher than in 1979, have brought down the average displacement of the boats and their stability and yet have given them more

The Spring 25 demonstrated Tony Castro's ability to break the mould of conventional thinking

volume and beam. I would have to conclude on reflection that the recommendations as far as I remember them, although by no means ignored, have had a limited effect.

Question: Has the emphasis on seaworthiness as a part of the design equation shifted over the last decade?
I would say yes it has but, as a result of the pressures to build different, faster kinds of boat, the emphasis is on designing for minimum acceptable standards, because of cost-related matters, and not necessarily on augmenting the safety factors across the whole range of boats.

In the process of designing new kinds of boats, the question of seaworthiness is continually addressed, but unless the brief calls for a particular emphasis on seaworthiness, the priorities tend to shift quickly to how fast, how cheap, how big is the galley, etc.

However, in particular on my production boats, in general the ballast ratios have gone *up* from 37-39 per cent to 40-43 per cent. The beam and volume have nevertheless increased as a result of commercial pressures. Only in the case of Barracuda were we allowed to take a completely fresh approach, substantially different from current production boats. Every report of that boat in bad weather, in gale force winds, has been full of praise for its handling characteristics. Some of it appeared in the press about one occasion when it was in the Irish Sea on a Force 9 gale doing 18 knots with triple-reefed main but neutral helm! Obviously some of those characteristics must surely be considered desirable.

Generally speaking as far as I am concerned, we are continuing to increase stability in all our designs, especially cruising boats, now fitted with wing keels of very low VCGs. Given the chance, we would also reduce beam, and opt for a moderate to light displacement. I think these fulfil the need to be fast passage-makers, with good handling characteristics and reasonably lightly loaded structures. I would always prefer an adequately strong, longer, lighter, less equipped boat to a smaller stuffed-up cruiser for the same price. The industry and public have something to learn in this respect, I believe.

I'm afraid we can only design in so many inexperience-proof features. Human beings will always be human, and experience and seamanship will always account for a very large percentage of safety at sea.

I would suggest that the biggest problem facing a boat today, inshore or offshore, is that it is being driven by the same people who expect and demand 120mph cars, when the speed limit is 70mph but, curiously, seem to be willing to accept the information from the speedo and rev counter. Everybody knows that if that needle is pushed into the red zone you can expect a broken engine and/or loss of your driving licence. Unfortunately, more and more people buy boats who don't have that minimum understanding and experience, that sense of a boat's limitations; without the equivalent of a rev counter, they don't know when limits are being reached...and, when they are, the consequences at sea are just as devastating.

Sailing steadily on through the storms of racing developments are the healthy, seaworthy products of the major production builders who have avoided the lure of the rating rules

Patrick Roach

CHUCK PAINE

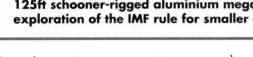

Chuck Paine and his team, whose headquarters are at Camden, Maine, USA, are major international designers of offshore sailing yachts whose Bowman 40 and Victoria Yachts from 26-34ft are well known. A designer of racing yachts for Dick Carter until a few years ago, his present interests range from a 125ft schooner-rigged aluminium mega-yacht to exploration of the IMF rule for smaller cruiser-racers.

Question: What design attributes go towards making a cruising boat suitable for offshore sailing?

I believe an offshore cruising yacht ought above all else to be *forgiving*. The term implies that some sin might exist which needs forgiving, and I believe that all sailors, regardless of the depth of their experience, commit 'sins' of greater or lesser severity. Be they errors of omission or commission, sailors being human will eventually be guilty of one or the other; and the boat they take to sea ought, by virtue of its design, to compensate for such errors insofar as designers can envision them.

'Forgivingness', to coin a word, can be divided into a number of clearly distinctive attributes. While all must be present, no one attribute of a design should overpower the others. For example, a designer might, in over emphasising stability, sacrifice steering qualities to the broad flat stern; worsen the motion with flat middle sections and a bow that pounds horribly; expose the deck to spray owing to low freeboard so that it is dangerously slippery in anything beyond a Force 5; and overballast to the extent that structural integrity is suspect. A proper offshore design not only incorporates all of the design attributes I am about to delineate, but does so in a harmonious blend such that the sum total, which I choose to call 'forgivingness', is maximised.

Watertightness: Surely the most important attribute of a serious offshore yacht is the ability to stay afloat. Even the worst offenders in the Fastnet débâcle did manage to stay on the surface, and resulted in loss of life for other reasons. Yet I fear that the design imperatives adrift in that other branch of yacht design, racing yachts, are treading very close to the edge of the abyss. I detect a rising attitude amongst that fraternity that yacht races are now conducted in sufficiently close proximity to rescue, or that liferafts have improved to such an extent, or that the challenge of winning is now so great that even mortal risks must be assumed. All such excuses lead to the construction of yachts that are structurally unable to withstand the sort of conditions that will eventually catch up with any yacht that is sailed well offshore. The solution is a simple one. Any yacht that is truly intended for offshore sailing should be designed to a recognised structural standard such as Lloyd's, ABS or Bureau Veritas. All of the yachts I have designed for construction in the UK have been done to Lloyd's scantlings, although only the Bowman is built under survey. Very few racing yachts, or production cruiser-racers, are designed to a recognised standard and, as such, they are bad risks for offshore cruising.

Stability: Second in importance only to the ability to stay afloat is the propensity to remain upright. Stability is a more difficult characteristic to enforce, though it is trivially simple to calculate. The designer of a dangerously

> **Even the worst offenders in the Fastnet débâcle did manage to stay on the surface, and resulted in loss of life for other reasons**

tender yacht can at least point at her on her mooring and say, 'There, look, she's upright.' Had she lost watertight integrity, there would be nothing for the embarrassed designer to point at!

A proper offshore sailing yacht should be sufficiently stable (definition to follow) without resorting to movable ballast. This means that her design stability should be figured *without crew sitting to windward*. In fact, even the assumption that the crew is situated on the centerline, evenly distributed between the cockpit and below decks (the assumption I use in my weight estimates), is scarcely a conservative one, for any bloke who has squeezed more that a cupful of salt water out of his socks knows that, after a few days offshore, it is the leeward bunks and cockpit seat that will be the prizes in those subtle games that crewmen play.

As for bodily ballast, so also for the liquid variety. I, among others, have been active in the design of water-ballasted yachts of late, a consequence of the great singlehanded races which have demonstrated the tremendous advantages to be gained from water ballast to windward. Yet there ought to be a standard to keep us all honest. That standard ought to be that water-ballasted yachts should be sufficiently stable (definition still coming) with an equal amount of water ballast in the windward and leeward tanks. Then

> **I fear that the design imperatives adrift in that other branch of yacht design, racing yachts, are treading very close to the edge of the abyss**

any additional stability, and speed, which results from filling the windward tank, represents simply an additional bonus.

So, an offshore cruising yacht numbers stability among her primary attributes. But how does one define sufficient stability? I do it in three ways. First, there exists in an American treatise on the subject of design, *Skene's Elements of Yacht Design*, a graph which relates stability to waterline length. Included are four curves, indicating 'normal' and 'tender' stability for both keelboats and centreboarders. I design any yacht intended for offshore service so that its stability falls on the 'stiff' side of the normal curve.

Secondly, I use a computer model that rotates the yacht through 180 degrees and prints out a full stability curve. Any design which fails to retain positive stability through at least 120 degrees of roll is subject to revision, or relegated to the 'coastal cruising' category.

Finally, there is a rule of thumb I call the 20-20 rule. In my view, any yacht that is to spend the greater part of its life offshore should be capable of carrying full sail in 20 knots of apparent wind (about 16 knots of true wind) without heeling more that 20 degrees.

13

Chuck Paine's Victoria 30

A potential offshore sailor is better off in a large yacht of indifferent design than an adroitly designed small one, for the simple reason that stability is related far more closely to size than shape

In my design work, I employ a very accurate VPP (Velocity Prediction Program) which predicts not only speed but heeling angle. Whenever I design a yacht that is in striking distance of the 20-20 target, I subject it to the usual iterative redesign process until that objective is met. This final requirement regrettably eliminates small boats as serious offshore cruising possibilities, as 20-20 is virtually unachievable in yachts of under 35ft in overall length. Yet I believe that a potential offshore sailor is better off in a large yacht of indifferent design than an adroitly designed small one, for the simple reason that stability is related far more closely to size than shape.

Controllability: It goes without saying that an offshore yacht must be controllable. I would extend the adjective a bit by the addition of a few conditions. An offshore yacht should be controllable by a single person (all others perhaps being incapacitated with seasickness) and that person should be assumed to be advanced in age and a bit infirm in both mind and body (the majority of my clients falling into this category).

Control begins with steering. One can get into a great deal of trouble very fast when a hull pressed over hard results in the boat rounding into the wind. Steering qualities are designed into a yacht; they are not a matter of black magic or luck. Proper offshore yachts should exhibit a reasonable balance between the bow and stern sections, and they should derive their stability from a low centre of gravity rather than flat sections. Yachts of this type are blissful steerers.

Racing yachts, in contrast, with their sharp bows and broad, flat sterns, are inherently prone to broaching with anything short of a professional helmsman in command. The rudder should be larger than the minimal (racing) size. It should be located as far aft as possible and of a proper airfoil shape (our office is constantly upgrading to the latest foils developed by the aerospace industry). The foil should be chosen for its forgiving stall characteristics rather than a marginal superiority of lift to drag ratio in steady state flow. Finally, the forces transmitted to the helm (that's you, folks!) should be minimised. This may be done by fitting a balanced rudder, with approximately 20 per cent of its area disposed forward of the shaft upon which it pivots. Or when the rudder is unbalanced, the quickness of the steering should be sacrificed to increased mechanical advantage in the gearing, yielding two to three wheel turns lock to lock. Racers are down around one turn or less.

A few additional control factors: Primary winches should be located within easy reach of the helmsman. The jib is the powerhouse of the modern yacht design's sailplan. Therefore its controllability is paramount. If the jib is fitted with roller furling, as it should be in this day and age, the helmsman should be able to wind in the furling line without taking both hands off the helm; simple if the furling line leads fair to the primaries, and they are within reach and self-tailing.

Finally, it should be possible to reduce sail quickly without leaving the security of the cockpit. The image of the oilskin-clad youth, strapped to the base of the mast, with the wind howling and the rain blown horizontal should be relegated to the Hollywood soundstage, yet this is standard reefing drill aboard many a yacht. Both in-the-mast roller systems and single lines-aft slab reefing exist today in workable form, and both are being continually improved. No yacht ought to go offshore without one or the other.

Motion: Offshore cruising yachts should have what is called an easy motion. Easy motion derives from a number of factors, including weight, stability, appendage geometry, and shape. Needless to say, heavy yachts have better motions, generally, than light ones. This is a consequence of Newton's first law of physics (F=MA). Given a certain force (a breaking wave, for example), an object of double the mass of weight will be accelerated exactly half as much.

Stability has an influence. Yachts with large GM (distance between their centre of gravity and metacentre) roll more quickly than those with smaller GMs. A 10 ton yacht with a GM of 2ft will have identical stability to a 5 ton yacht with a GM of 4ft. The difference will be that the yacht with the smaller GM will roll far less quickly.

Appendage geometry is important. The keel acts as a 'roll damper'. Any appendage (keel and rudder) that is extremely effective at preventing sideslip will also be effective at attenuating rolling. Thus the keel that enables a design to sail closest to the wind will have the added benefit of reducing rolling motion. This is a strong argument in favour of deep keels, although I shall shortly champion an opposing point of view. The practical alternative is one of the modern end-plated keels (wing keel or the proprietary Paine Keel, Collins Tandem Keel or Scheel Keel).

Finally, there is sectional shape. Yachts with round sections, the shape required on racing yachts to reduce wetted surface, tend to roll more easily and to greater amplitudes than U or V-shaped hulls. In fact, research conducted many years ago with the object of developing the ideal shape for bell and gong buoys, where it is desirable to produce the maximum rolling motion with the minimum wave input, produced a shape not unlike that of the present-day IOR racer!

Draught: I believe an offshore sailing yacht ought to possess *appropriate* draught. Clearly shoal draught is inappropriate for offshore work. Offshore yachts require serious stability, beyond the reach of shoal draughters. There are those who contend that in oceans whose depth may be measured in miles, there are no limits to the depth

Both in-the-mast roller systems and single line slab reefing exist today in workable form, and both are being continually improved. No yacht ought to go offshore without one or the other

of the keel. Would that life, or design, were so simple.

Prudent navigators realise that in the face of predicted heavy weather, discretion is the better part of valour. The famous ocean sailors, whose names are known to all of us, could recite by heart the names of the world's great hurricane holes. Almost invariably, these mangrove-rimmed havens are safe because they are protected by ramparts of shoal water. The bravest of these 20th century Nelsons would agree that, if true ocean-going stability (and roll damping) could be achieved at a draught midway between 'shoal' and 'deep', they would choose that option. And that objective is today within reach, through the use of one of the end-plated and end-weighted keels mentioned above. In the last four years, during which my office has designed ten major offshore yachts, every one has featured one of the moderate draught, low CG keel options.

Speed: Yes, speed is a necessary attribute of the offshore sailing yacht. Speed gives the captain the option of running for shelter when the need arises. A fast yacht, when unintentionally slowed down in heavy weather, is more comfortable than a slow yacht travelling closer to its hull speed. The option of travelling at good speed towards the chosen destination under reduced canvas and at a comfortable upright angle is part and parcel of 'forgivingness'.

Question: Have the 1979 Fastnet Race Report recommendations had any effect upon the direction cruising yacht design has taken since 1979?
Yes, certainly. In my own office I have added calculation of the capsize screen value to the initial suite of computer programs that are run on each preliminary design. Values greater than 2.00 raise a clear flag, and are altered.

But while design offices might alter their design processes slightly, it is the effect that the Fastnet disaster has had upon the clientele, and the designs they choose to commission, that will have the more far-reaching consequences. The history of yacht design seems to progress slowly, with isolated 'breakthroughs' in the form of individual yachts that satisfy a need, and sweep the whole field forward in occasional giant steps.

I submit as a potential example of such a step, my own design for an extremely knowledgeable client whose whole concept was to commission a reaction to the Fastnet-demonstrated excesses. This design was entirely driven by the Fastnet Report.

As I suggested, there can be no clearer evidence of the influence of the Fastnet results upon cruising yacht design than this particular yacht, which is now under construction.

Question: Has the emphasis upon seaworthiness as a part of the design equation shifted over the last decade?
Yes and no. Yes in the minds of certain designers and members of the sailing fraternity, as I hope I have illustrated.

No in the racing arena. In fact, the leading IOR yachts are more lightly constructed, have a greater proportion of their ballast inside, rely to a far greater extent upon crew for ballast, and have lighter spars and sails exacerbating capsize tendencies, than 10 years ago. Only the obvious shift from IOR to handicap rules more tolerant of lower VCG and narrower beam to length ratios (Channel Handicap and MHS) gives you hope that another Fastnet disaster will be averted amongst the 'other' branch of the sailing fraternity.

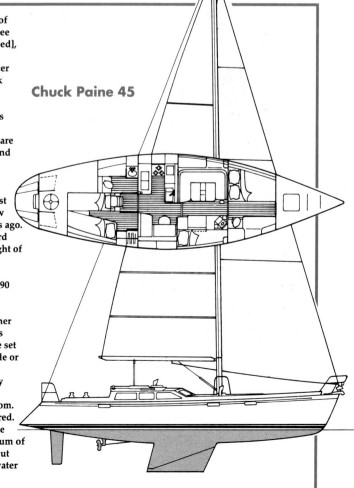

Chuck Paine 45

1. The beam to length ratio is moderate, and is complemented by a VCG of .342ft, *below* the waterplane. This gives a stability exceeding all my three targets (Skene's curves, 120 degree vanishing point [127 degrees achieved], 20-20 rule).
2. Stability above is calculated with no crew to windward and (fresh) water ballast tanks equalised. Pumping drinking water to the windward tank reduces the heel angle by 4 degrees.
3. Capsize screen value is 1.741, comfortably below 2.00.
4. Mast is oversized double spreader section significantly increasing mass moment of inertia for improved capsize resistance.
5. Structure is designed to a minimum of ABS standards. Certain aspects are done to multiples of ABS. Example: hull shell (of Airex cored S-glass and Kevlar in epoxy resin) has moments of inertia nearly double ABS requirements. Three interior bulkheads are watertight for virtual unsinkability.
6. Keel is state-of-the-art wing keel, designed by American aerodynamicist Dave Vacanti. Base keel and wing foils are modern NASA laminar flow sections, replacing the more common NACA sections designed 40 years ago. VCG of keel is 82 per cent of its depth below the canoe body! Windward ability and roll damping are comparable to a deep fin, but with a draught of 5ft 9in.
7. Sectional shape is a rounded U and does not change dramatically from amidships to stern. Change in trim with heel throughout roll from 0 to 90 degrees does not exceed 3 degrees. Steering will be sublime in all conditions.
8. Deck is designed for singlehanded control. Roller furling outer and inner jibs can be trimmed, reefed or furled from the helm. Mainsail reefing is single-line slab type, three reefs with full-length battens enhancing the set of the reefed sail. Runners can be permanently set up with either double or triple reef, boom swinging clear beneath them. Bulwark boards of epoxy-coated sitka spruce surround the deck, and are complemented by full-length jacklines.
9. Liferaft stowage is in a locker which opens outboard through the transom. Pulling a pin dumps liferaft on to scoop transom deck, no lifting required.
10. Speed. VPP predictions assure just shy of 10 knot speed in 20 knots true wind. Speeds of all points of sail in 7 to 10 knot range over wide spectrum of wind speeds and angles. All predictions are done at full liquid loans, but without credit for water ballast. Lighter loadings (half load) or use of water ballast will result in higher speeds still.

DAVID THOMAS........

David Thomas is an ex-Merchant Navy Master Mariner who worked as *Yachting World's* Assistant Editor for eight years, then Development Manager for Ratsey & Lapthorn. His first commercial design was the Elizabethan 31 in 1966 and he went on to draw the Elizabethan 30/9m, Quarto, and subsequently the Sonata, Impala, Medina, Horizon 26 and 32. He has designed all the Sigma range and is currently designing a 42ft steel cutter, as well as the 67ft steel cutters Chay Blyth is using for his British Steel circumnavigation project.

Some yachts are a joy to be at sea with. I was once caught at the helm of a beautiful wooden Cheverton *Danegeld* cruising yacht when the rest of the crew felt 'disinclined' to come on deck. With far too much sail up, we sailed each other for a whole day to Dinard and it turned into a total love affair.

She was designed 'right up to the gunwale'. In other words, her designer had taken as much trouble with her topsides as he had with her bottom. Reserve buoyancy was perfectly distributed in her ends and quarters so that she heeled and pitched comfortably and seemed to find her own way to windward with a minimum of help from her tiller. It was a joy to be alone with her. Many years later we belted across the same stretch of water in a boxy little Quarter Tonner called *Quarto* with the wind dead on the nose and four of us holding the little beast upright draped over the weather rail all night. She flew the course in winds that dropped an Admiral's Cup yacht's mast and arrived ahead of all manner of larger yachts. We were exhausted but it was the sail of a lifetime.

Which of these two yachts was the more seaworthy? They were both 'in a fit state to put to sea', which is the dictionary's description of the term, but one was cruising and the other was racing. The first yacht took care of her crew, took them to windward in spite of their shortcomings, whilst the second demanded every ounce of skill from the crew to prevent her from getting a wave wrong and being thrown on her beam ends. Both yachts achieved their designed objectives perfectly - but reverse the crews and it would have been a different story.

So we have the extremes of the offshore yacht and those who have a clear preference for racing *or* cruising have a simple choice, or have they? Cruising across open water can be achieved in every conceivable craft from Frank Dye's 16ft Wayfarer dinghy or the redoubtable Drascombe Lugger to the classic world-girdling *Wanderers* of the Hiscocks. Uffa Fox cruised in a sliding seat canoe and Dr Pye in a Falmouth Quay punt.

But on the basic subject of yacht design, it rather drew the conclusion that pretty well any yacht was at risk in these conditions

The urge to cruise and adventure on the sea should not be dampened by too long a search for the perfect cruising yacht. The ideal cruising yacht is a blend of yacht and owner in a perfect harmony of design, construction and skilful navigation. Even the Hiscocks had troubles with their choice of yachts.

The modern cruising yacht is the perfectly compromised and sometimes controversial production cruiser that sails calmly, with smiling crew, across the coloured adverts and brochures of every major yacht builder. She must perform equally as well at a Boat Show as she does on her brochure. Unfortunately, in terms of offshore sailing and the need to be able to take the crew to windward in heavy weather, some of them fall short. It is difficult for a prospective owner to identify easily a yacht's limitations. By 1992 we may well see a compulsory plaque in our production yachts describing their designed role, permitted crew and cruising range!

If we agree that the basic requirement of an offshore cruising yacht is the ability to sail to windward in gale force conditions then what are her likely characteristics? Ample lateral area in keel and rudder, moderation in freeboard and top hamper, power to carry enough sail to overcome the resistance of windage and waves when sailing to windward. Good equipment is needed to control the rig and strong hatches and windows to keep the water out. The ageing chestnut of whether the underwater profile should be continuous or separate fin and skeg has now been resolved in favour of fin and skeg. It is important to remember that a racing yacht, which is usually sailing at optimum speed, can manage with a small high lift, high aspect ratio fin of low lateral area. To sail slowly under full control, and to be able to leave the helm for short periods, a fin of probably twice the area is needed. The same applies to the rudder which can be too small, but not too big! Full depth rudder skegs are rather difficult to build in GRP and cause the production builder some problems when moving his hulls around the factory. However, I know of at least five lives that were saved by a skeg rudder, my own included, when we surfed under jury rig into Civita Nova, a concrete block-lined harbour entrance in a 30ft yacht. Two people striving on the helm just managed to avoid a disaster that an ordinary spade rudder might not have been able to cope with.

Our beneficial legacy from the currently waning period of the International Offshore Rule (IOR) has sadly not contributed much to the improvement of the offshore cruising yacht. During the 1970s it was fashionable to take the current Ton Cup Champion yacht and adapt it as the role model for the following year's production cruiser-racer. But not any more. The IOR has been responsible for a marked

One of David Thomas's first boats, the Elizabethan 31 has become a classic cruising design

Above: the Sigma 33 class had a rough baptism in the 1979 Fastnet but it was the foundation of bigger sisters, the 362 shown here and the 38 introduced in 1988. Left: although controversial, Thomas's Quarto design was successful if a handful to sail well

increase in speed potential for a given waterline length, but paid for in terms of low inherent stability, a flat bottom, lack of bilges and twitchy directional control due to minimal lateral area. Great beam at deck level and ample freeboard were offset by decreasing waterline beam leaving the power to carry sail dependent on more and more crew on the windward rail. The yachts were prone to rolling and broaching and carbon fibre and Kevlar began to creep into more and more expensive construction.

In its first page statement on Rule Management Policy the IOR states:

'The Council (Offshore Racing Council) will act to discourage developments which lead to excessive costs or reduce safety or the suitability of yachts for cruising.'

This sentence has long brought a smile to anyone who spotted it at the beginning of the rule book.

The disastrous Fastnet Race 1979 did little to help matters. Conditions were so diabolical that yachts suffered from a new hazard seldom encountered in our coastal waters, wave collision. Normally, yachts sail around in bad weather crashing into seas. In the Irish Sea in 1979, waves crashed into yachts, threw them on their beam ends (the B1 knockdown) and, too often, into the complete B2 capsize. Keels, instead of acting as pendulums, started to act as flywheels. Some beamy yachts when tossed into the 180 degree inverted state became quasi-stable. The principles of Static Stability on which the IOR relied were of little use when hydrostatics turned into hydrodynamics.

The Fastnet Race Report produced valuable recommendations on race organisation, search and rescue systems and procedures. It highlighted specific weaknesses in steering gear, companionways, storm sails, liferafts, lifejackets, engines and electrics but on the basic subject of yacht design it drew the conclusion that pretty well any yacht was at risk in these conditions and placed the onus firmly in the hands of the Offshore Racing Council 'with a view to their considering whether further changes in the measurement rules might not be required.'

When speed is the main criterion, seaworthiness passes largely into the hands of the crew

An incident like the Fastnet Race gives everyone concerned in the yachting industry cause to look at their products. Those who were actually taking part in the event lost little time in rectifying the shortcomings of their products.

There was not much time for the ORC to gather its wits between the publication of the Fastnet Report in late August and their Annual Meeting in November. They attempted to guide designers to increase the stability of their designs but as usual they were hampered in their efforts by the need to avoid upsetting the existing fleet and its Ton Cup rating bands. The surge of developments towards the 'Offshore Racing Dinghy' continued to run ahead of the Council and its Technical Committee. IOR devised a new rule as an interim measure, the Channel Handicap System.

This was a simplified rule to hold weekend racers together, mainly in production boats; just what was needed. It was harsh with 'grand prix' yachts that tried to sneak in for easy pickings and, most important, it did not discourage stability. The Royal Ocean Racing Club formulated a Stability Screening System to weed out some of the smaller yachts unsuitable for offshore sailing. Cruiser-racers, old and new, started to appear on the coastwise race courses. Production yacht builders heaved sighs of relief and began to think about healthier yachts. Designers started to straighten their buttocks and increase stability and God-fearing IOR owners were seen sailing with the CHS fleets or even moving into the two successful family cruiser-racer One-Designs, the Contessa 32 and Sigma 33.

The lunatic racing fringe began to take interest in racing around the oceans in maxi-raters, singlehanded water-ballasters and multihulls.

Meanwhile 170 cruising yachts sailed across the Atlantic in the 1988 ARC event from the Canaries to Barbados. Only one yacht came to grief; fortunately her crew, who had not expected to arrive in Barbados so quickly, were saved.

Sailing steadily on through the storms of racing developments are the healthy, seaworthy products of builders like Rival, Westerly and Moody who have avoided the lure of the rating rules, only adopting the occasional healthy titbit from the competition table. There is a continuing increase in the number of yachts designed for shorthanded cruising with steady hulls that can give a sailing 'couple' more time to be seaworthy. Timing is the essence of good seamanship.

More and more cruising owners are gaining the confidence to venture offshore with the help of better training and improved communications, both with the yacht through electronic monitoring and with the shore through navigational aids and RT, and with the help of improvements in sail and ground tackle handling equipment.

The moral of our story seems to be: 'Never rely on a yacht built to a Rating Rule as a guarantee of seaworthiness. For when speed is the main criterion, seaworthiness passes largely into the hands of the crew.

Seaworthiness in cruising terms must be firmly in the design and construction of the yacht!

Cruising yachtsmen have grown into a wily breed, easily as clever as their racing counterparts. Cruising yachts call equally on the skills of seamanship. Their big advantage is that they do not have to keep changing their rules.

Chapter 3

Watertight integrity

Hull and deck openings are the greatest threat to a yacht's overall seaworthiness. This chapter looks at good practice and practical improvements that can be made

THE WATERTIGHT INTEGRITY of a yacht's hull lies at the heart of its seaworthiness and capability as an offshore yacht. While small leaks can be dismissed as annoying, major leaks or failures of part of the hull structure or fittings can demoralise the crew, distract them from the smooth and efficient running of the yacht, and eventually lead to its loss.

Ideally it should be possible for a skipper to seal up his yacht in such a way that it should be impossible for water to get in short of catastrophic damage to the hull itself. A buoyant, sealed container must remain afloat as long as it remains unruptured.

The best way of achieving this has taxed many minds over many years. It has been a prime concern of those who organise offshore yacht racing, particularly the Offshore Racing Council, and their conclusions are enshrined in the booklet *Special Regulations Governing Offshore Racing*. We will be referring to this publication many times in this book because we believe its requirements form an excellent model when setting up an offshore cruising yacht.

The ORC is the governing body for offshore racing world-wide and races under its auspices range from low-key, inshore events under handicap, level rating and class rules to high profile, long-distance events for top flight yachts and crews. Clearly requirements appropriate for a transatlantic race would be inappropriate for a club regatta for a local one-design class, so they divide their requirements into a number of categories from 0, the most stringent and designed for trans-oceanic racing, to 4 which covers races held during daylight in protected waters.

We have chosen as the model for offshore cruising Category 2. This covers design, construction and equipment for yachts on 'races of extended duration along or not far removed from shorelines or in large unprotected bays or lakes where a high degree of self-sufficiency is required of the yachts but with a reasonable probability that outside assistance could be called upon for aid in the event of serious emergency'. In almost all respects the requirement is the same as for Category 1, in which the yacht is expected to be able to fend for itself without outside assistance well offshore and in all weathers. The ORC requirements are paraphrased right.

The dilemma facing a yachtsman tackling the question of the watertightness of his yacht is: which openings are only likely to produce irritating minor leaks, which have

1 Check mast gaiters for cracks
2 Fit waterproof deck plugs
3 Big hatch tops are liable to carry away
4 Cockpit drains should clear flood water in 3 minutes
5 Hatch boards must remain in place through 180 degrees
6 Seacocks must be double clipped
7 Windows should have storm boards for offshore use
8 Dorade boxes should have sealing caps
9 Crew must shut hatches at sea
10 Anchor/stowage wells must drain properly

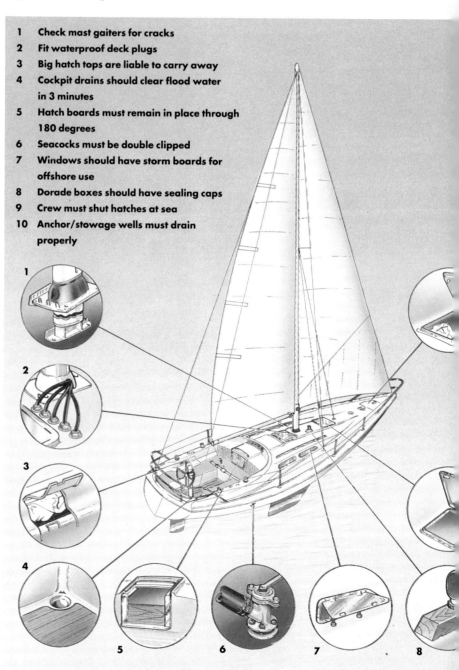

the potential in time or under extreme conditions to imperil the yacht, and which are capable of causing the total loss of the yacht under normal sailing conditions? Those which pose no threat at all under normal conditions may be highly vulnerable to extremes, while others which are likely trouble-makers, even when the yacht is sitting at her mooring, may be no more likely to cause loss in a hurricane.

In looking at all the weak points in a yacht, we have rated each area from 0 (no risk) to 5 (high risk), first according to its risk in good sailing conditions and then *in extremis*. Thus an inward opening port in the forecabin might be rated 2/4.

Hull structure (0/5). Modern GRP hulls are, for the most part, well built and adequately

A breaking wave is always the greatest threat to a yacht. It sweeps the deck looking for every weakness and hits vertical surfaces with devastating force

Peter Haward

strong. Structural failure of the hull was not a factor in the Fastnet disaster, and cases of yachts being lost because the hull has failed through the action of wind and wave are rare. Sadly though, loss through holing following collision with floating debris has been increasing due to the indiscriminate dumping of flotsam into the sea.

There is little, in practical terms, that an owner can do retrospectively about his hull. Indeed, for most coastal and offshore cruising little that is practical is needed. At the building stage, though, reinforcement of the bow area can be specified or a yacht chosen which has been designed to be 'unsinkable' by incorporating foam between inner and outer mouldings.

Yachtsmen venturing far offshore might consider building in one or more watertight compartments. The obvious area to seal off is the forepeak or forecabin. This will prevent water from a hole in the bow flooding

ORC Special Regulations Category 2

This is an outline of the requirements for hull integrity and watertightness with the paragraph numbers referring to the paragraphs in the Regulations. Copies of the booklet are available from the ORC, 19 St James's Place, London SW1A 1NN (price £2.00, plus 20p p&p).

The essence of the regulations are contained in paragraph 6.1. 'The hull, including deck, coachroof and all other parts, shall form an integral, essentially watertight, unit and any openings in it shall be capable of being immediately secured to maintain this integrity'.

Hatches are a major concern of the booklet, reflecting the report on the 1979 Fastnet disaster which pointed to the main hatch as one of the most vulnerable parts of the yacht. Paragraph 6.12 states: 'No hatch forward of Bmax (maximum beam) shall open inwards except ports having an area of less than 110sq in (710cm²). Hatches shall be so arranged as to be above water when the hull is heeled 90 degrees. All hatches shall remain firmly shut in the event of a 180 degree capsize. The main companionway hatch shall be fitted with a strong securing arrangement which shall be operable from above and below (6.12).

The report adds that all washboard and hatchboards shall be capable of being secured in position with the hatch open or shut and shall be secured to the yacht by lanyard or other mechanical means to prevent their being lost overboard (6.13).

The companionway should not be extended below main deck level ideally, but if it is, it must be capable of being blocked off to the level of the sheerline abreast the opening. Such a blocking piece shall not prevent access below (6.14).

Cockpit design comes in for a lot of attention. It must obviously be essentially watertight and all openings must be capable of being securely closed. The report treats anchor wells and any other deck well as a cockpit for the purpose of the requirements (6.21).

The requirements stipulate certain maximum dimensions for the cockpit (6.23). While it is usually impractical to change the size of a production cruiser's cockpit, an idea of what is considered sensible is no bad thing. The maximum volume (below the lowest coaming) shall be 9 per cent of the load waterline x maximum beam x freeboard abreast the cockpit (9 per cent L x B x FA) and the cockpit sole must be 2 per cent of the LOA above the loaded waterline.

The requirements for cockpit drains (6.31) are complicated, but essentially yachts over 28ft should have drains the equivalent of four threequarter inch drains (allowing for screens), while yachts under 28ft

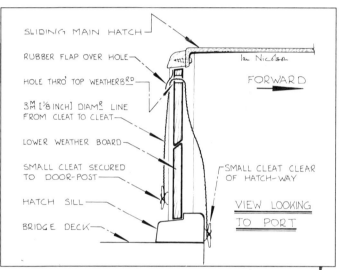

Above: an ingeniously simple method of securing the companionway from inside or out. Right: a line of windows like this is in contradiction of the spirit of the ORC Regulations

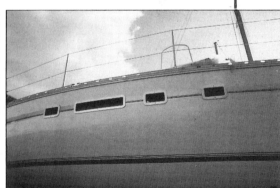

Patrick Roach

should have the equivalent of two 1in drains. All drains must work at all angles of heel.

Other requirements of the rule include storm coverings for all windows of more than 2sq ft (6.4) and seacocks or valves for all through hull openings below the waterline except for integral scuppers, shaft log speed indicators and the like, although these should have some means of closing the opening when necessary (6.51). In addition, softwood plugs of the correct size should be carried attached to, or adjacent to, the relevant fitting (6.52).

Bilge pumps (a minimum of two) are required to be so fixed that one is operable above and the other below decks with all cockpit seats, hatches and companionways shut (8.21.1). Bilge pumps should not exit into the cockpit. Handles should be attached to the pump by a lanyard (8.21.4) and finally, on the basis that the best pump is a frightened man with a bucket, two buckets with lanyards should be carried (8.24).

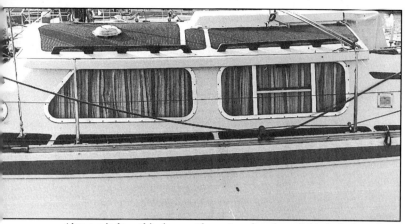

Above: windows this size must have storm boards available for bad weather.
Right: Ian Nicolson's answer for leaking hatches, particularly in cockpit wells

the rest of the boat and may provide sufficient buoyancy to keep the yacht afloat, if not sailable, should a hole develop aft of the bulkhead.

On the subject of unsinkable hulls, Bill Anderson of the RYA commented that possibly the greatest benefit was in the confidence of the crew. They would be less likely to abandon the yacht for the liferaft, preferring instead to stay with the yacht, effect repairs and make for port. In the Fastnet disaster seven of the fifteen deaths occurred away from the yachts and six were lost after being washed overboard. Only two died on board their yacht.

To cope with holing when it occurs, the well-found offshore yacht should carry at least a selection of softwood plugs, which can be used to staunch small leaks, and a canvas sheet (or storm jib) to spread over the outside of the hull in the event of major holing. In addition, thought should be given to how such things as bunk boards, cushions and other loose items could be used to staunch leaks. Yacht delivery skipper Peter Haward recommends carrying a bag of quick-drying cement to give a watertight seal after stemming the worst of the flow by other means.

Keels (0/4). It is rare for a keel to fall off a boat, whether as a result of collision, grounding or simple structural failure. More common are leaks in way of the keelbolts, caused either by grounding/collision or deterioration of the bolts themselves.

There is little, retrospectively, that an owner can do to reduce the risk of keel damage, but a prudent buyer will look at the attachment carefully with a particular eye for large backing pads, substantial floors and large bolts. Glassed-in bolts corrode more slowly and are less likely to leak in the short term, but in the event of damage they are a nuisance.

Should a keel fall off in any strength of wind, you will be lucky to save the boat. Leaking through cracks or via the bolt holes can be treated like hull leaks, but get to shelter as soon as possible because a loose keel will quickly work looser and fall off. Take all possible action to reduce the motion of the boat, particularly pitching and slamming.

IN EMERGENCY

What to do when a major leak develops

Collision damage
When the hull is ruptured after a collision, finding the leak is not usually the problem. Slowing, then stopping, the flow is. Proceed as follows:
1 Tack to put the damage as close to the waterline as possible
2 Take steps to stem the flow (see above)
3 If water level continues to rise, summon assistance and prepare liferaft
   Note: the priority is to keep the boat afloat. Only divert attention from this to seeking aid when it is clear the yacht will sink

Sudden and unexplained flooding of bilge
1 Taste the water. Fresh water indicates a leak in a water tank or the engine fresh water coolant
2 Operate electric bilge pump, put crew on to manual pump. Reduce water level by all means possible - it will make leak tracing much easier
3 Check all seacocks. If loose reseal or remove (have your softwood plug handy). Shut off if pipework is faulty
4 Check anchor well/chain locker for blocked drain or flooding
5 Shut down engine. This will reveal an exhaust coolant water leak
6 Check stern gland or outdrive seal
7 Check keel bolts
8 Check bilges for signs of water flowing down hull sides. Ignore trickles, concentrate on major flows. Trace flow back to source
9 Check for hull damage or spring of planks/plates. Plug as required

Companionway/main hatch (1/5). The Fastnet report identified the companionway as potentially the most vulnerable part of the yacht in extreme weather conditions and, in particular, in the event of a B1 or B2 knockdown (B1 means a knockdown to the horizontal, B2 means knockdowns to substantially beyond 90 degrees including full inversion).

There are conflicting design requirements involved in the companionway. For most of the time yachtsmen want one which is wide and deep with a low bridgedeck for easy access, washboards which are light and quickly removed, and a hatch which is also light in weight and transparency. In practice this can be combined with a design which will be more than adequate for normal sailing. But in storm conditions it will be highly vulnerable. Washboards must be capable of being secured in place, the whole must be capable of being sealed up and opened both from below and from the cockpit, the bridgedeck must be high and of a reasonable width. A single washboard is more secure than two or three. A simple method for securing washboards to the boat is shown in the diagram on page 91. Some production yachts now have hatches which can be operated from below and on deck, but proprietary systems, such as the Sea Sure Fastnet Latch, are available to owners wishing to make the modification. (Tel Sea Sure, 0489 885401)

Peter Haward suggests that the sliding main hatch should be replaced with an aluminium frame type deck hatch whose after edge beds down on the top washboard.

Windows, opening and fixed ports (2/4). Bill Anderson has this to say on the subject: 'These were not an extensive cause of trouble in the Fastnet, but I suspect it would be a different story today if a fleet of production yachts were caught at sea in a similar storm. I have no reason to believe there is anything wrong with the design and construction of the fittings, it is just that there are now so many more deck hatches per boat. Any hole is a potential source of water ingress, if only because once in a while someone won't shut the hatch properly.' Designer and surveyor Ian Nicolson describes windows as 'the most vulnerable part of the hull shell'.

The ORC recommendations are clear and sensible yet more and more production yachts, many intended for Mediterranean sailing, are in clear breach of the spirit and

How flooding affects a yacht's stability

When we asked the Wolfson Unit for Marine Technology at Southampton University for their comments on watertight integrity, they provided some revealing information about the effects on stability of comparatively minor flooding of a hull. These are shown in the accompanying diagrams.

They also highlighted the speed with which water can enter a hull through apparently small openings. Here are some examples: A yacht under spinnaker pinned down by a gust could take in through a 100mm vent in the side of the coachroof 1 tonne (220 gallons) of water a minute if the vent was just 0.5m (1ft 7in) below the water. A 600mm square hatch 0.3m below the water could take in a tonne in just 1.5 seconds.

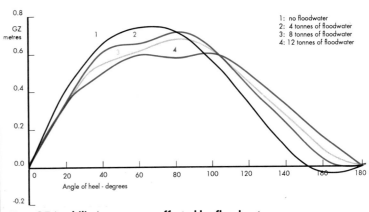

How GZ (stability) curves are affected by floodwater

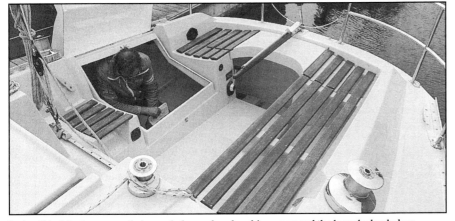

Variation of draught with quantity of floodwater 11.6m (38ft) yacht of 7.4 tonnes displacement

of flood on the waterline acht

An ORC approved 710cm² opening port could let in a tonne in six seconds if forced open and submerged to 0.5m.

The Wolfson Unit concludes, 'The figures illustrate the importance of locating all openings close to the centreline of the yacht and ensuring that they remain securely closed in adverse conditions - and that means the closure cannot be separated from the opening'.

Leaks below the waterline do not have to be large to be catastrophic. A crack just 1mm wide and 200mm long, located 1m below the surface, will let in a tonne in 20 minutes. 'This might not sound a lot until you are the one who has to bale it out at the same rate for a prolonged period'.

The Wolfson Unit points out that not only does a significant amount of flood water dampen the spirits of the crew; it also reduces speed, freeboard and stability. Additional openings can become submerged and waves sweep the decks endangering vent cowlings and deck hatches.

The question of stability is an interesting one. For most yachts stability is not greatly affected because as the yacht sinks its beam and form stability increase. But sailing angle will be increased thus reducing performance and, in particular, ability to make to windward. The Unit points out that, since floodwater also reduces inverted stability, in the event of a knockdown flooding can help to right the yacht.

Locker lids this size can easily be carried away by a breaking wave and the large locker below flooded, considerably damaging the yacht;s sailing ability at best, sinking it at worse

letter of the 'law'.

Again, in practical terms there is little an owner can do to reduce excessive areas of glass, but he can take action to minimise its vulnerability. A rigid code of practice for the crew in battening down before leaving harbour is a good starting point. The next step is to provide storm boards for each opening. The purpose of these is to protect the windows during a storm, but they should also be capable of being used in the event of a window being stove in.

Ian Nicolson has given us this advice for storm boards. Ideally the board should be screwed to the hull or superstructure and bedded in mastic, but in practice it is sufficient to arrange the board so that it drops into runners each end of the window, rests on a shelf below it and is held in place by turnbuckles along the top edge. The board should be made of half-inch ply with a stiffener across it for larger panels. Mark each board with top, bottom, fore and aft. Large, wheelhouse type windows, he says, should be protected every time the yacht puts to sea, though most yachtsmen would find this impractical. Holes can be drilled in the board to allow light into the deckhouse.

Although wood is the traditional material for storm boards, polycarbonate may be stronger and has the added bonus of letting in light.

Other recommendations for strengthening windows include 'double glazing' by adding an oversized sheet of polycarbonate to the outside and fitting wood or steel battens horizontally or vertically to the inside so that they touch the glazing.

Deck hatches (2/4). Modern deck hatches are structurally sound and should present few problems short of minor leaks as the seals deteriorate with age. The important thing is to remember to close hatches before leaving harbour. Peter Haward recommends that hatches should have levers for securing them inside and out.

Older GRP hatches can cause problems, though, as Ian Nicolson points out. Seals perish and peel off, hinges and clasps become weak and break. He has known them to just slide overboard or drop into the bilge. In addition, the GRP itself becomes brittle and cracks.

The answer is to re-glass it on the inside, bolt on new, oversize hinges and fit man-sized clamps. New sealing strips should be added both to the coaming and to the hatch edge so that water is trying to pass two seals. Finally, through-bolt two 3in by 1½in grabrails/stiffeners along the top.

Replacing an old wood or glassfibre hatch with a modern aluminium framed one can be complicated by the need to provide a flat bed on a curved coachroof. The answer is to build up the coamings with wood.

Cockpits (1/5). The security of the companionway and the size of cockpit drains is dealt with extensively under the ORC Special Regulations and the Fastnet report (see special panels), but that still leaves the vexed question of cockpit lockers and hatches.

Few boats are designed today with engine hatches in the cockpit sole. They are a potential disaster. At best they weep salt water on to the engine and any ancillary electrical systems. At worst they can fall off in a B2 knockdown and lead to the loss of the yacht. They should be sealed off as permanently as practical.

Ian Nicolson recommends bedding the hatch down on a flexible sealant such as Farocaulk or a tape sealant such as Inseal, in which case mitre corners and scarf-join the ends as shown in the diagram. The hatch should be through-bolted down on to the seal using 5mm bolts at 4in centres and reinforced as necessary.

Once secured, the hatch should not be removed except for major engine overhaul or replacement, after which it will need resealing.

Alternative access to the engine can be provided (often much more satisfactorily) by cutting side hatches by the quarterberth and in the cockpit locker, making sure to retain the integrity of the soundproofing.

Cockpit locker hatches are almost equally vulnerable. This is what Bill Anderson says on the subject. 'I can't think why we ever had cockpit lockers. Almost by definition they are cavernous spaces with tiny openings at

Fastnet Report

The document which has come to be regarded as the definitive study of offshore seakeeping, the *1979 Fastnet Race Inquiry Report*, deals at lenght with the question of watertight integrity, and made many recommendations which are now included in the ORC Special Regulations.

The report identified the main hatch as the most vulnerable area, and its recommendations concerning securing the washboards and sealing the hatch from above and below are now part of the ORC requirements.

A yacht abandoned during the 1979 Fastnet storm still afloat despite the open main

The report also highlighted the fact that washboards in veed companionways could fall out if lifted only a short way and advised that companionways should be straight sided. This suggestion has not been followed up. The ORC has also failed to adopt the report's advice in introducing a time element into the design of cockpit drains. The report recommended a maximum draining time of three minutes.

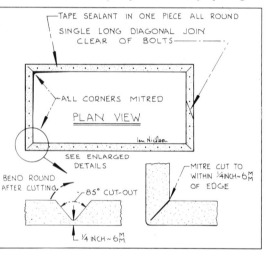

The correct way to fit tape sealant to a cockpit locker

the top. They have an enormous potential for becoming the apocryphal Midshipman's Sea Chest - everything on top and nothing handy. And I don't think I have come across a locker lid which is really watertight.

'The fashion for cabins under the cockpit has helped to get rid of cockpit lockers. This is definitely an improvement in design. It gives good sea bunks in the right part of the boat, nobody has to clamber past sleeping occupants, and the motion is the least violent. The next obvious stage is to leave fo'c'sles empty for stowage.'

An ideal solution used to be found in older yachts which had a lazarette for general stowage aft of a watertight bulkhead. With access via an aluminium-framed hatch with the handles uppermost, this would be convenient and secure.

Locker lids are, though, an inevitable fact of life. A good lid is small, does not break the line of the coaming, has deep drainage channels down each side, has three or more stout hinges and at least two catches. The locker below it should be sealed from the main part of the boat, be shallow rather than deep and compartmentalised to make organisation of the contents possible. The seals on the locker lid should be checked regularly.

Mast gaiters and deck plugs (1/1). Keel-stepped masts frequently leak at the apertures and the only solution is regular re- placement and adjustment of the gaiter. Deck plugs are also a regular source of leaks. Installing good quality plugs such as the Bowdeck, or a swan neck type pipe stopped with a plug of silicone sealant, will go a long way to solving the problem.

Stern glands and outdrive seals (1/1[4]). Another source of irritating leaks, but seldom a source of danger, is the stern gland. Saildrive seals very rarely fail, but when they do the result can be catastrophic. It should be checked regularly and maintained/replaced in accordance with the manufacturer's instructions. Stern gland greasers should be tightened regularly (once per hour of engine time) and the packing replaced from time to time. Modern sealed glands and bearings are generally more leak-resistant.

Skin fittings (0/3). These should be fitted with proper seacocks, not gatevalves, greased regularly and checked periodically to ensure they remain securely fixed to the hull. Pipework should be double-clipped **at both ends.** Once again, good seagoing discipline in shutting cocks can save a lot of trouble later on.

The engine cooling water inlet seacock is particularly important here. Between the inlet and exhaust the water travels through a maze of pipework, any one part of which is vulnerable and may become unclipped. Apart from seacock discipline, annual maintenance **must** include a full check of all pipework and clips, replacing and double clipping if in doubt.

Exhaust pipes (1/3). A good installation will have a high swan neck in the piping, a syphon break and a water trap. On some boats (mainly multihulls) the run may be straight from the engine to the skin fitting. If it is impossible to add a swan neck, the alternative is to fit a shutter to the exit to prevent water splashing or syphoning back into the engine.

Remember to have softwood plugs of the correct size secured near all skin fittings including transducers.

Ventilators (1/3). Most modern ventilators are perfectly capable of withstanding the

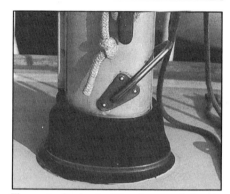

The mast gaiter is a traditional source of chronic leaks. Although usually minor, they are irritating and damage crew moral

Right: which openings are only likely to produce irritating minor leaks and which have the potential under extreme conditions to imperil the yacht?

onslaught of moderately severe conditions, but for gales there should be blanking off caps to seal them. Dorade vents on cambered decks must have drain holes on both the downhill and uphill sides to allow for the heel of the yacht.

Bill Anderson comments that there are not enough ventilators on modern yachts. 'The tendency to fit more opening hatches has been an excuse for a lack of sufficient ventilators.'

Crew discipline

On putting to sea: close all hatches and windows; close seacocks; check bilge pumps; clear strum boxes; turn automatic pumps to manual (otherwise they may conceal a developing leak); clear cockpit drains, deck well drains, scuppers; secure storm boards and watertight bulkheads as necessary.

Chapter 4

The seaworthy rig

It seems that long past are the arguments over how many masts a yacht should have and what proportion they have to each other. Aboard the modern offshore yacht the single mast rules, but detail is all

WHEN TAKING the magnifying glass to the modern offshore yacht, it is in the rig and rigging that significant advances, rather than necessarily changes, have been made. For the cruising boat, materials haven't altered very much in the last 15 years but the design of rigging hardware has allowed innovative new systems which in turn are beginning to influence rig proportion and configuration.

Refinement through innovation and detail has been the theme rather than any major changes. One of the advantages of this process of evolution is that older boats can be kept up to date with the latest developments.

The single greatest influence on cruising rigs over the last 15 years has been the roller reefing headsail which is now fitted to an estimated four out of five modern cruising yachts. Its advantages need no repetition here, but a roller headsail nevertheless has two principal disadvantages; one, no single sail serves all conditions a yacht is likely to meet in the normal course of cruising; and, two, their shape when reefed can be poor aerodynamically.

These factors have, in turn, influenced the rig configuration of modern boats because at the same time as the cruising headsail has gone through the roller revolution, the mainsail has also made great strides. Initially, these came with the wide introduction of well-designed slab reefing and, more recently, the popularisation of full length battens, lazyjacks and the leading of all control lines aft to the cockpit. The net result is a mainsail that is significantly easier to control and handle than a sail of the same area would have been 10-15 years ago.

The logical marriage of the roller headsail's disadvantages and the mainsail's improved handling has resulted in a two-pronged move, on the one hand, to bigger more powerful mains (re-creating the genuine *main*-sail) which demand less emphasis on the flexibility of headsail size and, on the other more conservative hand, to new popularity for the cutter rig.

Generally speaking, the bigger mainsail configuration comes in the form of the fractional rig. It tends to have a racing background, and is therefore treated with an amount of suspicion by the cruising fraternity, not all of it reasonably justified. On the plus side, the fractional rig has a number of advantages:

1. The headsails and spinnaker are considerably smaller in area. This makes them easier to handle for the weaker crew, and allows savings by using lighter gear.
2. There is much less emphasis on the headsail as a driving force, so the foresail wardrobe can be minimised, cutting costs and stowage space below.
3. The shorter forestay is easier to keep tight therefore increasing windward ability and the efficiency of roller gears.
4. Compared to the masthead rig, fractional is a better rig downwind without using a spinnaker or cruising chute – its big main pulling well compared to the smaller, masthead mainsail which not only provides worse drive but blankets the powerful headsail.
5. The small overlap on the headsail allows the main to be eased out without stalling. Combined with a fully-battened sail this decreases leeway and heeling forces, reduces weather helm and is a desirable feature in blustery conditions.
6. Although the average cruising yachtsman may not take too much advantage of it, the fractional rig allows the possibility of bending the mast to flatten the sail and thereby reduce the frequency of reefing.

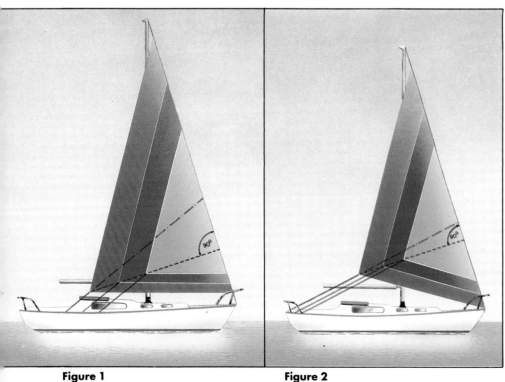

Figure 1   **Figure 2**

A good example of a modern cutter, the Victoria 34, with all the attributes of an up-to-date rig

Left: sail design for cruising yachts has improved beyond recognition in recent years - two examples are the fully battened main and the radially cut roller jib

Figure 3

a large foresail because the staysail makes up the area.

In this way, the progression of reefing (largely without leaving the cockpit) is an easy one, rolling away an efficiently set foresail gradually, reefing the main to preserve balance and keeping the staysail set throughout. One is left with the snug inboard rig of staysail and double reefed main, which is reduced further by reefing the staysail (either rolling or slab-reefing) down to storm jib size. Ultimately the double reefed main is substituted with a trysail.

By comparison, the masthead rig is simple, robust, has lighter loadings and will withstand all manner of abuse which, for all its warts, inevitably makes it a continuing favourite in the cruising fraternity.

The roller headsail

As mentioned above, the roller headsail has revolutionised the cruising man's lot. Over

Patrick Roach

Against it, from a cruising point of view, is the fact that, usually, it either requires swept back spreaders which are difficult to tune and far from ideal for mainsail trim and chafe during offwind sailing, or the complication of runners which by concensus are bad news aboard the average cruising boat. Figure 3, however, offers a viable alternative.

Fractional rig also demands a particularly well-balanced hull with predictable and reliable steering/tracking abilities, because off the wind the big mainsail exerts its centre of effort a good distance away from the yacht's centreline.

The cutter is making something of a comeback for a number of reasons. First, on boats over 40ft the cutter helps to keep headsails to a sensible size.

Secondly, Figures 1 & 2 demonstrate that the more equal leech and foot lengths are (ie more Yankee-shaped), the better a sail will roll, and the less its sheeting angle needs to be changed. Greater angle between mast and headstay encourages this sail shape and the cutter rig which, with its mast a little further aft, tends to have this, doesn't rely on

Buoyant masts and their effect on ultimate stability

The Australian yachting journalist Andrew Bray (not YM's Editor) recently wrote an interesting article in his magazine *The Cruising Skipper* outlining the thinking behind having a buoyant mast in a knockdown situation. Inspired by a SNAME (USA's Society of Naval Architects and Marine Engineers) report about stability in the light of the 1979 Fastnet, he came up with some startling results for his own centreboard boat.

The Wolfson Unit did similar calculations for us. Taking data from a 30ft cruising yacht, whose mast dimensions were known, they added the mast to the computer definition of the yacht and recalculated the stability to obtain an accurate result.

The diagram above presents their findings. It includes the effect of 50kg of water held in the mast and free to move along the mast as dictated by the heel angle. Fifty kilograms represents about 50 per cent of the mast's total capacity. This water reduces the stability when upright because it raises the VCG (Vertical Centre of Gravity) and reduces the buoyancy of the mast when inverted.

If the owner of an existing yacht wants to increase its range of stability at no extra cost, how about hoisting a large fender or two to the masthead on a spare halyard when the going gets rough (the fenders would need to be attached in a way that didn't put all the strain on their normal lanyard attachment points which would probably break under strain)? The effect of such action is also shown on the diagram, assuming just one fender 9in in diameter and 27in long (1cu ft of buoyancy) and weighing 5 lb is hoisted with its upper end at the masthead. Two such fenders would apparently yield almost the same gain as a watertight mast with none of the inconvenience or expense.

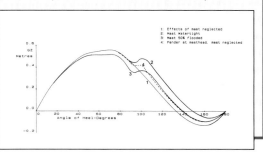

the years, the hardware has improved in detail and a variety of ideas have come and gone. The double swivel (swivels attaching head and tack of the sails), generally fitted to top-of-the-range gears, allows the centre of the sail to roll before the top and bottom, therefore taking the fullness out of the centre of the foresail. This improves the set, although a significant advance has been made with the introduction of a tapered foam-filled luff pocket inserted just inches abaft the luff tape. This induced bulk improves the sail's set through a good proportion of its rolling range, and has the advantage that it can be retro-fitted to your existing sail.

When ordering a new sail for a roller gear, one must take into account several factors. No single sail covers all conditions and the offshore yacht should be equipped, ideally, with three headsails (although very often they have two). These comprise a ghoster, the working rolling headsail, and a storm jib (Figure 4). Only the working sail should roll, the others set complete or not at all.

One effective arrangement has the ghoster set loose luffed on a lightweight furling (not reefing) gear immediately ahead of the principal forestay. It is furled as one tacks, reset on the new heading, and is stowed and handled as a rolled sausage. In overlap conditions, where the wind varies between needing the ghoster and the primary roller headsail, it can be left hoisted, but rolled up.

Size and proportions of the working rolling headsail are very important if the sail is to be a success. Too large a sail will be OK in lighter conditions, but on the assumption that, rolled beyond 50 per cent area it will be unacceptably inefficient, there will be a big gap between it and the storm jib. The *raisons d'être* of a roller reefing sail are convenience and avoiding the need to leave the cockpit, and yet many a rolling sail is shaped in such as way that the sheet track has to be adjusted constantly to maintain efficiency.

The ideal roller headsail for a sloop is a high-clewed 140 per cent (of foretriangle area) sail, something like a No2 genoa but with clew raised. It appears that the ideal cut is a radial shape, either single or bi-radial (for yachts 35ft plus) and, if YM's experience in boat testing is anything to go by, this is certainly our favourite. The disadvantage is that the sail doesn't lend itself to having a sun-strip added so needs either to be lowered after use, covered with a sock (hoisted on a spinnaker halyard) or the sail constructed in UV-resistant cloth. Bearing in mind the importance of size and shaping (Figure 2), W G Lucas, amongst other sailmakers, recommend traditional mitre-cut rolling headsails if the boat is to be used in a sunny climate so a sacrificial sun-strip can be employed.

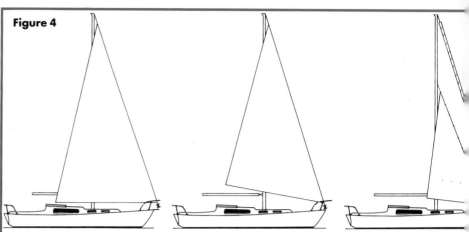

Cruising inventory for a yacht with a primary roller headsail system. Left, light ghoster, perhaps set flying; centre, the roller working sail with highish clew; right, storm jib set on a separate stay

Having a vertically-cut headsail with heavier cloth weights in the after panels (which are exposed in heavier conditions) is good in theory, although in practice it tends to create a sail that sags in lighter going, and easily hooks along the leech.

A final point about roller headsails relates to their use. Anyone who has seen how well one of the new generation of in-mast roller mainsails sets will realise that no sag in the leading edge is a large part of the answer to setting problems for roller headsails. The forestay *must* be taut if the sail is to roll well, and this will rely on an easily-adjusted backstay. If you're nervous about keeping this tight with the boat under way, at least tighten it up temporarily when reefing the headsail, and don't leave too much wind in the sail (thereby bending the forestay) whilst reefing. To this end, a roller whose control line is winched in under load will not only stress the roller equipment but is more than likely to create a poorly set sail to boot.

The modern mainsail

The refinement of slab reefing systems through well-designed reefing booms improved the efficiency of the average sailing cruiser considerably, not only by making reefing quicker and easier, but by creating a well set sail when compared to roller reefing booms of yore.

In gale conditions, a deeply rolled working headsail is never a substitute for the dedicated storm jib set up on an inner forestay

One detail worth pointing out to those considering conversion of their roller boom to slab is to make sure that the reefing pennant starts attached to the boom 2-3in (50-75mm) aft of the reef position, goes up

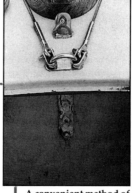

A convenient method of attaching an inner storm jib stay is by passing a chainplate through the deck and through - bolting to the anchor locker bulkhead. This one leaks, although it is not a significant problem when the water goes into the self-draining locker

An excellent feature seen on many French boats is the fitting of a lanyard to the front edge of the headsail sheet car and led back to the cockpit, thereby eliminating the need to leave the cockpit when reefing a headsail

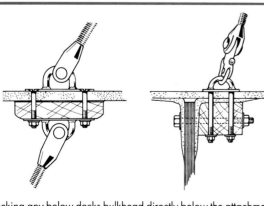

Lacking any below decks bulkhead directly below the attachment point for a storm jib stay, either a wire span can be taken down to a strong point on the hull or it can be braced to a nearby bulkhead

Leading of lines aft

Any system of leading lines aft is going to create a good deal of friction and this can easily defeat one's first attempt at this emminently practical and seamanlike arrangement. The halyard turns through 180 degrees at the masthead which obviously cannot be avoided, and the 90 degrees at the mast step is also inevitable. However, as far as is practicable, try to minimise subsequent turns on the line. The above photo shows good and bad - the 90 degree turn induced by the right-hand set of halyard organisers is completely unnecessary when one sees the shallow angle their sisters in the left-hand strip are making.

High quality, over-sized blocks are by far the best way to go when leading lines aft, even if you use standard ones elsewhere in the rig.

The highly popular Bamar in-mast mainsail reefing can be retro-fitted to an existing conventional rig. As can be seen here, the set of the sail is excellent throughout the rolling range

through the cringle *but then leads to cheek blocks fastened to the side and far end of the boom.* If the cheek blocks are situated near the reef position, the bunt of the reefed sail is badly pinched and damage invariably results.

Over the years, deck hardware has improved steadily, with well-priced roller bearing blocks and in particular the development of jammers and clutches which revolutionise line handling and enable many lines to be taken to a single winch. These developments have led to the logical extension to a slab reefing system of leading all control lines aft.

Because all boats are different, there are no commercial kits available to convert one's boat to lines aft, but it's rarely beyond the averagely practical yachtsmen to do it himself, bearing in mind the heavy loads that some of the lines, in particular the halyard and reef pennants, come under.

If your mast is keel stepped, make certain that the deck around the mast aperture is structurally capable of withstanding the upward pull of control lines. A full study of the various methods of rigging lines aft was published in YM in April 1988.

Whilst the mainsail has been tamed when set, various ideas have been appearing on methods of controlling it as the sail is lowered, broadly divided into commercially-available products and the home-spun lazyjack (See Figure 5).

In the commercial field, for conventional mainsails, a popular and economic option is the Dutchman, a simple system of vertical wires hung spanning topping lift and boom,

A concentration of drilled holes (particularly large ones) around the mast base like this is asking for trouble

passing in and out through eyelets in the sail. It works well (even better with fully-battened mainsails), holding the bunt of a slab reefed sail and helped along, flaking the mainsail as it lowers.

Another cruising development which is becoming very popular in mainsails is full battens. The fully battened mainsail offers many advantages, not least of which is easy stowing, especially if lazyjacks are used. The weight of the battens, and the fact they hold the sail out evenly along its horizontal length, makes the sail drop easily and stack itself like a junk's sail. Another major advantage, and this needs to be experienced to be appreciated, is that the sail does not flog which creates an air of calm during sail handling which is good with a slightly nervous crew. This aspect results in another of the fully battened mainsail's advantages - prolonged life of approximately 30 per cent because of no flapping and the fact that the full battens distribute the stresses evenly around the sail. Looking ahead, this increased life expectancy will justify the extra 10-20 per cent the sail will cost.

Other plus factors for the fully battened mainsail are its great efficiency for motor-sailing (even when you appear to be head to wind, it is still providing lift and steadying power) and it lends itself more readily to a single-line reefing system. The sail will also outperform a soft mainsail.

Against the fully battened mainsail is principally friction and resultant chafe, especially around the batten ends, so the sail will require more regular valeting. Much of this can be avoided if the sail supports only a modest roach, so compression within the battens is kept to a minimum.

Vertically rolling mainsails are justifiably gaining great popularity aboard offshore yachts. Initially their reliability, and yachtsmen's conservatism, threw something of a question mark over them, but these concerns are fading into the past. Convenience and ease of handling make using any of these systems a joy.

Convenience aside, the principal advantage of roller mainsails is their ability to tame the sail completely with two results. One, it allows a yachtsman to sail a bigger boat than he might otherwise consider. Secondly, for the yachtsman whose age is starting to limit his activity, or whose young family crew have flown the nest, fitting a roller mainsail allows him to handle his boat for many years longer than he might otherwise expect.

Using roller mainsails couldn't be easier; the efficiency of the very flat sail is excellent in heavy weather, allowing one to reef infinitely and, unlike the headsail, it maintains its shape until the last. The only danger is the temptation to over-sheet the sail when heavily reefed which will stretch the leech badly, or at worse, rip the sail. *If a roller sail (either main or headsail) rips, never under any circumstances roll it away - remove it from the spars there and then.*

A mainsail will lose about 10 per cent of

Lazyjack refinements

Take the lines out along the spreader a few inches (to minimise batten snagging), leave enough fall so that sailcover setting is simplified by easing the lazyjacks and hooking them around an object in the gooseneck area. Finally, the system only works satisfactorily if the lines come to the top edge of the boom. If you do not have slug slides on the boom, pass the lines through an eyelet on the foot but, so the system is easily adjusted (and dismantled for sail removal), pass the lines through either side of the eyelet and figure-of-eight knot both on each side.

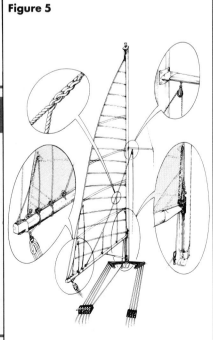

Figure 5

its area (from the leech and foot) by being fitted to a roller, so ideally mast height should be increased marginally to accommodate this. The ideal roller system is the dedicated in-mast equipment fitted to the yacht as original equipment, which allows one to fit a slightly taller mast. However, the new breed of roller designed to rivet on the after face of an existing mast are very good, and other than losing some mainsail sail area, do all that the dedicated in-mast furlers do, although not quite as neatly aesthetically.

The only real disadvantages of in-mast gears are weight aloft, which obviously reduces stability and increases the boat's motion in a seaway, and the inaccessibility of the 'works' inside the mast or extrusion. Assuming stability isn't a particular problem with the boat rigged normally, the extra weight is unlikely to be dangerous.

Downwind sails

The preponderance of big headsail/small main designs in the seventies put greater emphasis on downwind running sails and towards the end of that decade was born the cruising chute, under a variety of names. In theory, it was a sail which had many of the advantages of the spinnaker without the latter's complexity and equipment. In practice, on boats of 30ft plus anyway, the cruising chute needs nearly all the equipment of the spinnaker - a pole (if it is to be used on a broad reach or downwind) plus uphauls/downhauls and some form of squeezer - but comes nowhere near the performance of the spinnaker. Neither can it be flown over as wide a range of wind angles.

The broad concensus is that the spinnaker is the better of the two sails, but used and rigged as a cruising sail. For cruising, the sail will be flown with the apparent wind well aft of the beam in conditions of no more than Force 3-4. Within these parameters, it should cause no problems. For a single sail used for cruising the best cut is radial head which is more stable off the wind (and some 20 per cent cheaper than its tri-radial sister), and ideally a little smaller than the IOR-designed spinnaker for your particular boat. Don't be tempted by heavier material because it's a cruising sail, it won't fill in the light conditions it will be most used in.

Chainplate design

For peace of mind, chainplates should look strong. The cross-section of metal should be at least four times that of the shroud, and four bolts is the minimum.

Because chainplates always move slightly there must be some arrangement at the deck to stop water seeping through. A plate screwed down on to the deck is common but, as in so many locations, bolting is better here. A bedding material like Farocaulk works well, as it is rubbery and can absorb the movement of the deck.

Most chainplate bolts have ordinary washers and it is common to see crushed wood where the bolt has been tightened up, perhaps several times in the course of a decade. A long plate washer is much better, because ample area of metal cannot be squeezed into the face of the ply.

The majority of bulkheads in yachts are made of 12mm plywood which is roughly ½in thick. This is fine for most normal cruising boats pottering along the coast. For ocean cruising and long range sailing, some doubling is advisable, even if it is only an extra sheet of 8mm (5/16in).

In terms of hardware, a single boom and single sheet/guys (used with a barber hauler) are quite acceptable for cruising, a spee squeezer of some sort taming the sail like a lamb for hoisting, lowering and, importantly, gybing.

Standing rigging

The $64,000 question about stainless steel 1/19 standing rigging is, how long does it last? Certainly, if it is going to fail through bad manufacture, it will do so in a matter of months. If you've a fitting that has always seemed slightly dodgy, perhaps a swaged terminal with a slight bend in it, the chances are that, if it is going to fail at all, it will have done so already.

The prospect of visually-intact shrouds snapping like carrots is more than worrying when you know you are pressing hard in bad conditions and every component of the rig is working to the limit. There is no simple answer, other than regular biannual close inspections for any signs of single strand failures (they're most common at terminals), and routine replacement of standing rigging every 8-10 years, dependent on the yacht's usage.

From an insurance point of view, there is no published data on when or not they will pay out on broken masts due to the age of rigging. Knox Johnston Insurance quote paying out recently for a 20-year-old rig

Out of line T-ball terminals can set up stress which may lead to early failure

which had broken, but we have also heard of cases where claims have not been paid 'due to the age of the rigging' when the rig is less than 10 years old. It goes without saying that every part of the standing rigging should be allowed to move universally at its two connection points, with toggles being imperative.

David Thomas recommends that, when buying second-hand, the yacht's standing rigging should be changed routinely, partly because you don't know how often they've been used to make fast springs and so on.

The thickness of standing rigging inevitably has a bearing on its lifespan - an oversized wire is under less stress than one working close to its limit. Having said that, beware of rushing out and fitting oversize rigging because it needs considerably greater tension to impose tight which can set up intolerable strains in the hull (trying, for example, to make a forestay tight), always assuming the mast doesn't crumple at sea through undesigned compression loadings.

Clearly the greatest source of problems is the component nature of a yacht's rig. With literally dozens of connections, the failure of any one of which could lead to losing the mast, care and good design in the initial setting up followed by diligence and regular

Rigging screws

It is a great pity more people don't throw away their rigging screws more often. Fewer masts would tumble down each season and we would pay smaller insurance premiums. A rigging screw is a moving part and as such it is not reliable. It is much more likely to fail than the wire of the rigging, or a chainplate. If

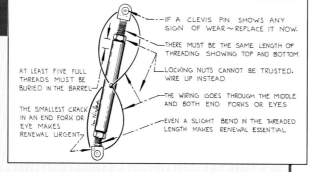

one of yours is old, if the screwed part is sloppy in the barrel, if the threaded part is even the tiniest bit bent, if there is a crack anywhere...be brave...dig into the bank balance and renew all.

In the notebook where you keep all the information about the boat, write the details of the rigging screws. Include the length of the threaded part, measured most accurately so that, when fitting out, you can check that one of the rigging screws is not hanging on by the last two threads.

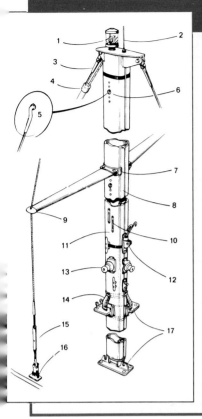

Rig check

1. Check masthead light, clean and grease terminals
2. Check aerials and other electronic sensors
3. Check upper terminals and anchorages, make sure blocks and sheaves are free and lubricated
4. Check roller reefing gear - is the halyard swivel free to rotate?
5. Check shroud terminals for alignment and cracking
6. Check shroud tangs and keyholes for damage
7. Check spreader roots for compression failure and security
8. Check mast track for smooth running and secure fastenings
9. Check spreader tips for chafe and cover ends to protect sails
10. Check sheaves for free running and wear
11. Check mast walls for pitting, chafe from ropes and halyards, impact damage and electrolytic corrosion
12. Check gooseneck fitting and lubricate
13. Check operation of all fittings, particularly winches, and service as necessary
14. Check mast gaiter for leaks and condition
15. Check bottlescrews for alignment, cracking and kinking. Are terminals in good condition?
16. Check chainplates for lifting and other signs of movement or damage. Are toggles adequate?
17. Check heel fittings and deck plates for distortion and damage

inspection are the only answer. Not only that, but problems are more likely to crop up at sea than in port so, despite modern rigs which aim to keep you firmly in the cockpit, never relax on constant inspections at sea.

Roller forestays deserve special attention as furling gears create a new set of problems. For example, depending on the design of your furler, a stiff (or worse, seized) upper bearing in the roller gear can unlay the construction of your forestay which will not withstand such abuse for long.

Another specific problem is the strain that can be imposed on the lower clevis pin of a roller gear. The regular and sustained torque action on it can, over a short period of time, have the effect of 'worrying' the pin sideways. This is not a problem if the pin comes up against its domed head, but if the natural inclination is to work up against the split pin, this sheering force will, in time, break the pin off, with disastrous results. Check this carefully and consider fitting a same-size nut and bolt (secured with a split pin or lock-nut).

One of the by-products of maintenance-free modern materials is the fact that yachtsmen often forget to use lubricants, which every moving part of the rig needs in the environment it works in. There are some excellent aerosol Teflon products for mast slides and blocks which will last much longer than lighter oils such as WD40. Good old anhydrous lanolin is still by far the best treatment for static equipment in the rig like bottlescrews and clevis pins.

T-ball fittings first came into use during the seventies when frequent un-rigging was called for on dayboats and their advantages (low windage, light weight, simplicity) quickly saw them introduced on to larger craft. Manufacturing problems in the seventies and early eighties resulted in a number failing at the neck, but that problem is now overcome. Boat with T-balls eight years old plus should replace them. Good alignment is imperative. The conservative yachtsman still tends to prefer tangs, being able to see and inspect his rig, but the probability of component failure increases. As a broad rule of thumb, standing rigging should not meet the mast at anything less than 12 degrees on a cruising yacht (10 degrees on racing craft).

In this day and age, the integrity of the chainplate is becoming ever more important as modern builders very often bring all caps and lowers to this single point for simplicity and economy. Other than putting all your eggs in one basket, it lays greater onus on the builder to distribute all the rig's windward loadings through the hull evenly and fairly.

Running rigging

The technology introduced to modern rope-making has transformed the lot of the cruising sailor, making the lines he uses less susceptible to chafe and breakage, not to mention having excellent low stretch characteristics for running rigging. To this end, many yachtsmen prefer to use braided rope for halyards, the stretch being minimal, which also enables them to end for end the line from time to time (making certain that snap shackles, etc are attached with a tight seized bowline rather than splice to aid this, and that halyards are all over-length to accommodate being trimmed back for chafe).

The reliability of braided tails and wire to rope splices have thankfully rung the death knell for captive reel winches (many of which were dangerous) and combination wire/rope halyards are very common. They have the easy handling advantages of rope, but the strength and low stretch of wire, which seems ideal but chafe can be a problem with the splice itself if it passes through sheaves regularly under much strain. Ideally, it should be placed in a position where the masthead sheave is the only one it passes over.

The argument between stainless steel and galvanised halyards appears to be six of one, half a dozen of the other. Galvanised is more flexible (requiring sheaves of sixteen times its diameter as opposed to twenty times for 7x19 stainless steel) but stainless more prone to fatigue (although this it does visibly by producing sharp barbs). Cosmetically, galvanised wire is a problem after a period of time due to rusting because the central core is rope and withholds moisture which degrades the wire.

Kevlar is a rope construction that has the potential to replace wire in halyards, weighing and stretching less than 7x19 wire. However, Kevlar is severely weakened when passed around even a normal-sized sheave, or when tied in a knot. With the resultant need for large sheaves sixteen times the rope's diameter) and special terminals, its practicality for the average cruiser is minimal until its weaknesses have been resolved.

Kicking straps are an important part of the cruising boat (especially with fully-battened mainsails), although as often as not they are fitted as an afterthought. Not only should the tackle be either powerful or able to be led to a winch, but great care should be taken with precise axis alignment of mainboom gooseneck and lower kicker fitting or damaging strains can be imposed when the boom is eased out.

The alternative, gaining popularity, is the rod-kicker which serves the extra function of obviating the need for a topping lift. The cost is, of course, greater than a block and tackle but the power advantages combined with lack of chafe from a topping lift are well worth it for the cruising yachtsman.

Split pins

Aboard the modern yacht the split pin has largely replaced seizing wire. To avoid failure, the split pin requires to be understood because it is easily abused. The table below states correct sizes which should be used, both length and thickness are important. Split pins should be replaced on a regular basis, every three or four uses, should never be bent more than 10 degrees out of straight (or 20 degree apart) and bound with tape to protect the ends.

The right and wrong method of securing split pins – never bend them back on themselves

Clevis diameter	Pin diam	Pin length
1/4	3/32	3/8
3/8	1/8	9/16
1/2	5/32	3/4
3/4	3/16	1 1/8
1	7/32	1 1/2

Chapter 5

Decks – the working platform

The deck and cockpit of a yacht are the shopfloor of sailing. It is from here that the crew work the yacht and spend much of their time both at sea and in harbour. Although safety and security of the working platform are of prime importance, there are other significant areas to consider

LIKE SO MANY parts of a sailing yacht, the deck is a compromise with elements that sometimes oppose one another in function and sometimes complement. The crew below deck demand headroom that results in a coachroof that has the added benefit of improving ultimate stability should the yacht be knocked down. That same crew on deck find that coachroof an obstacle when working and a nuisance when trying to find somewhere comfortable to sunbathe. Large deck hatches make for a bright interior and turn into skating rinks when wet. A razor sharp non-skid surface will be a godsend when changing a headsail in heavy seas but you'll have less charitable thoughts about it when you've got grazed knees and oilskins torn to shreds.

Cockpits

Cockpit dimensions. Whilst on smaller boats its physical size is dictated by the space available, on boats of over 30ft this ceases to be a limiting factor and there is no reason why the cockpit should not be an efficient working area for the number of crew on board. Any shortcomings in this area are more often than not due to the dictates of what is becoming one of the primary design requirements, the accommodation.

> **1979 Fastnet Race Inquiry Report**
>
> **These extracts, referring to deck equipment, are taken from the Official Report. This was based on questionnaires returned from 235 yachts**
>
> - There were six reports of broken harness attachment points. Some of these led to the loss of crew
> - Thirty-eight boats commented that there were insufficient hand-holds and safety harness attachment points
> - 126 boats had no special provision for attaching the helmsman's safety harness
> - Many of those who commented favourably on harnesses felt that two lines, each with its own hook, were an advantage
> - Twenty-seven boats said that non-skid surfaces were inadequate
> - Twenty-six reported that a surfeit of halyard falls and control lines was a problem
> - Forty-four lost items of deck equipment overboard
> - Forty-five lost items of distress/rescue equipment
> - Fourteen reported that items of distress/rescue equipment were too securely stowed
> - An RNLI coxswain commented on the lack of towing points forward
> - Only fifty-one boats had jackstays fitted

There are high, shallow, centre cockpits to allow for below decks walk-through passages; shallow cockpits to allow space above aftercabin bunks and cockpits with near deck height bridgedecks to give more standing headroom. The modern fashion for stern cabins combined with hulls that have shallow after sections both conspire to make a cockpit that is often far from ideal.

The disadvantages of a shallow cockpit are **(a)** crew security, **(b)** less shelter and **(c)** a cockpit that is more difficult to work, for example when using sheet winches (see illustration). Peter Haward, in stating his preference for deep cockpits, also points out one of their dangers, that of larger, more vulnerable companionways. Perhaps one advantage of the shallow cockpit is that, in the event of being filled, it holds less water and will empty more rapidly as the water spills out.

The relationship between cockpit width and depth are important. A deep well needs to be wider than a shallow one to make it possible to stand comfortably when the boat is heeled but not so wide that it is impossible for crew to be able to brace their legs against the opposite seat. Central fore and aft footbars are one solution, though the advantage they offer at sea may be outweighed by the inconvenience they cause in harbour. The well should not be so deep that the helmsman is unable to see over the coachroof and the same goes for the seats.

There should be adequate space for all a passage-making crew to sit in the cockpit and still allow the boat to be worked properly. Ian Nicolson suggests that the well should be long enough for one person (the standby watchkeeper in bad weather) to lie down in and still leave space for the helmsman's feet. Seats, too, should be long and wide enough for someone, perhaps a seasick crew, to lie down on without getting in the way.

High coamings are one solution to shallow cockpits as they give increased shelter, good back support and provide a better

> **ORC Special Regulations, Category 2**
>
> **We have chosen these regulations, which are safety and equipment requirements for offshore racing yachts, as a model, which is for yachts taking part in races 'of extended duration along or not far removed from the shoreline or in large unprotected bays or lakes where a high degree of self-sufficiency is required but with a reasonable probability that outside assistance could be called upon for aid in the event of serious emergency'.**
>
> **Below is a précis of the requirements for decks and deck equipment with the paragraph numbers referring to those in the Regulations. Copies of the booklet are available from the ORC, 19 St James's Place, London SW1A 1NN (price £2.00, plus 20p p&p).**
>
> - All heavy items...including anchors and chains shall be securely fastened so as to remain in position should the yacht be capsized 180 degrees (5.4)
> - Cockpits shall be structurally strong, self-draining and permanently incorporated as an integral part of the hull (6.21)
> - Where wire lifelines are required, they shall be multi-strand steel wire. A taut lanyard of synthetic rope may be used to secure lifelines, provided that when in position its length does not exceed 4in (6.61.1)
> - Pulpits and stanchions shall be securely attached. When there are sockets or studs, these shall be through-bolted, bonded or welded (6.61.4)
> - Taut upper lifelines, with upper lifeline of wire at a height of not less than 2ft above the working deck to be permanently supported at intervals of not more than 7ft (6.62.1)
> - Wire jackstays must be fitted on deck, port and starboard of the yacht's centre line to provide secure attachments for safety harnesses. Jackstays must be attached to through-bolted or welded deck plates, or other suitable and strong anchorages
> - Through-bolted or welded anchorage points, or other suitable and strong anchorage, for safety harnesses must be provided adjacent to stations such as the helm, sheet winches and masts (6.65)

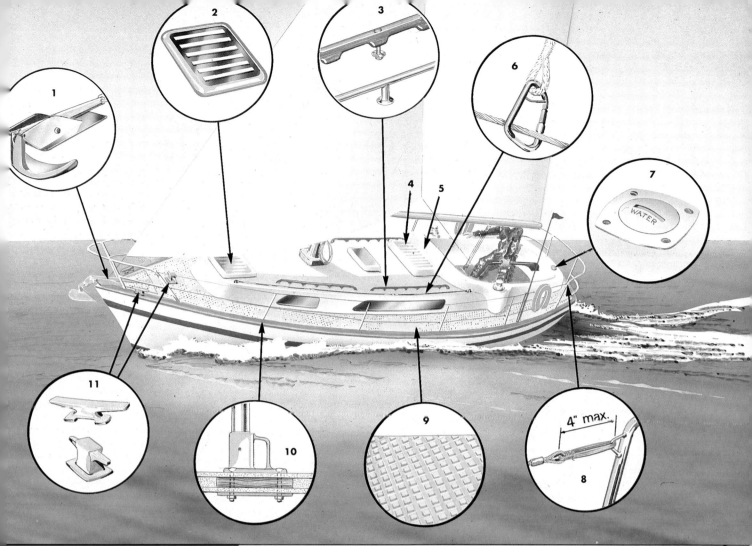

The offshore shop floor

1. Anchors stowed at the stemhead are convenient and labour-saving, even if their weight detracts from windward performance. They must, however, be properly secured, but beware drilling through the shank of a cast anchor

2. Deck hatches are like ice when wet. Self-adhesive non-slip strips are the answer

3. Grabrails must be through-bolted and have adequate clearance below. The best rails are of stainless steel raised several inches above the deck so they fall more readily to hand

4. Another skating rink, the hatch garage is rarely non-slip and should be treated the same way as deck hatches

5. The coaming is not just there for looks or a convenient attachment for the sprayhood. In wet conditions it will prevent water on deck sluicing into the cockpit or worse, belowdecks

6. If practical, the best site for jackstays is along the coachroof sides, as it is further inboard and therefore less likely to be walked on and roll underfoot. Jackstay strong points must be through-bolted and the stay fitted with proper terminals

7. Water and fuel fillers on deck can leak, leading to contamination. Coamings make a good site

8. Lanyards for securing guardwires offer several advantages. They are easily tensioned, provide an insulation break (breaks the round deck electrical loop) and can be cut easily in an emergency. Seven complete turns of 2mm Terylene is equivalent in strength to 5mm stainless steel wire

9. Non-slip surfaces should be a balance between effectiveness and the damage they cause to crew's knees and oilskins. This need not mean a trade-off in performance

10. Stanchion bases must be beyond question, through-bolted, reinforced and with backing plates

11. On most standard boats cleats are rarely up to more than marina berthing and moderate weather anchoring. If sailing on exposed coasts and anchorages, these will need to be at least a size larger and properly reinforced

winch platform (see illustration over) but if they are too high they become an obstacle for anyone leaving the cockpit to go on deck.

A bridgedeck or means of closing off the companionway to cockpit seat height is a necessity but higher is not necessarily better. The 'up and over' full height bridgedeck might, at first sight, appear the most secure of all until you consider the vulnerability of crew as they go below or come on deck. Such an arrangement completely separates cabin and cockpit, making communication difficult, and the height of the climb up the companionway is a major inconvenience.

Sheet winches mounted far outboard on wide coamings are difficult to use when heeled, not least of all because they may be under water. A crew member kneeling on a cockpit seat has to lean out to use them; he is working at a severe mechanical disadvantage and is dangerously unbalanced.

Convenience and comfort. Detail is important for cockpit comfort. Recessed cleats and engine controls are not just cosmetic. The horn of a cleat in your back in the small hours might help you stay awake but it is a major aggravation – as can be a protruding gear lever that snags every loose line and stuffs itself up your trouser leg as you move around the cockpit.

The fashion to cram more accommodation into shorter hulls means that cockpit lockers are now relegated into what space is left over, invariably right at the stern which, in terms of weight, is exactly where you don't want them.

Whilst stern lockers are valuable, there should also be stowage provided further forward for heavy objects – kedge anchors, outboards and tenders for example. Deep and undivided lockers are another bad feature because, unless they are well organised and subdivided, they end up as a huge glory hole with the one thing you need right at the bottom. Stowage for smaller items of working gear is necessary, be they odd lengths of line or the fishing tackle, and for such items cockpit cave lockers are excellent and possible to retro-fit. (Yacht Tidy, £22.50, from Sheraton Cabinet, White Oak Green, Hailey, Witney, Oxon, Tel: 099 386275.)

Peter Haward dislikes wheel steering on boats which physically don't require it because 'it leaves the helmsman stuck away in splendid isolation, little more switched on than an autopilot unable to help with tacking'. This is a point which Bill Anderson also makes, saying that on boats of under 35ft the mainsheet must lead to the helmsman as such yachts are often lightly crewed.

Sprayhoods are one of the simplest and most effective ways to improve the quality of life on board. Not only do they keep the cockpit crew out of the wet and wind but they also prevent water going below and allow the boat to be sailed in most conditions with the main hatch open, improving access and ventilation. The only disadvantages to consider are loss of visibility forward as the windows are rarely more than translucent, the fact that they may obstruct the step from cockpit to sidedeck and access to the companionway, particularly with high bridgedecks, and windage.

Dodgers are, perhaps, less effective when sailing to windward, although they do offer some shelter from the wind when on a reach. Their inherent windage can be a problem as can their tendency to scoop up water on the lee sidedeck which may damage both dodger and stanchions. A dodger should have a 6in gap underneath and either be laced with shockcord or attached with Velcro to prevent such damage.

Developments in sail handling for cruising mean more lines leading aft to the cockpit for reefing headsails and remote mainsail reefing. Whilst this is a major convenience and decreases the risks to crew working on deck, both Bill Anderson and Peter Haward dislike the resulting mess of lines with inadequate stowage.

If lines are led aft, they must be properly organised and there is little point having reefing pennants and main halyard led to the cockpit if you still need to go on deck to hook on the luff cringles. A system should be arranged so that one crew member can carry out a sail reduction from one station, whether at the mast or in the cockpit. The

Although deep cockpits hold more water, they offer greater security when seated (top) and a better working position when using sheet winches (above)

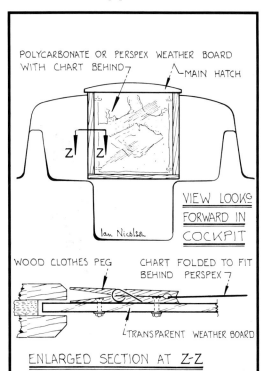

A transparent washboard can get over the problem of charts in wet and windy cockpits. Two or three wooden clothes pegs are bolted down the sides on the inside and used to clip the chart on. When not in position in fine weather, the board can be used in the cockpit on the navigator's knee

Although on the shallow side, there are a lot of good features here: the 'drain' could hardly be simpler and more effective, straight through the transom (netting would prevent small children and toys following suit); the seats and backrests are comfortably angled with recessed cleats; the sidedeck is angled for easy sitting out; gutters on the outboard edge of the cockpit seats will keep the leeward one dry; there are no cockpit locker lids; the bar on the inboard edge of the seats is good both for hand and toe holds

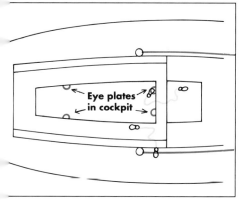

Cockpits should have harness eyes for both crew and helmsman. They should be sited so that it is possible to clip on before coming from below and jackstays close to hand to attach lifelines before going on deck

same goes for conventional headsail changing where the halyard should be at the mast so one crew member can control the lowering or hoisting of the sail.

Peter Haward comments that, although remote sail handling systems reduce the need to go on deck, they never remove it and labour-saving devices rob the yachtsman of the ever present necessity to be active. The more we stagnate in the cockpit, he says, the more we put the quality of seamanship at risk.

Cockpit security. The non-slip surface on the cockpit sole should be every bit as good as on deck, perhaps better as it is likely that drinks and food may be spilt making it slippery. Likewise, the areas where crew normally tread when leaving or entering the cockpit should be suitably treated, perhaps

using 3M non-skid strips on the coaming top and seats if smooth.

But the main line of defence for crew security in the cockpit remains the safety harness. The cockpit feels secure and so the temptation is to wear a harness but not to clip on except in extreme conditions, when a guardwire or mainsheet might suffice. Whilst crew probably are relatively secure when seated, nobody sits immobile for hours on end. There is constant movement, leaning to leeward to check for dangers, going below or on deck and, one of the most vulnerable moments of all for men, relieving themselves over the side. During any of these movements in fresh or strong weather it needs only a slight lurch of the boat to throw you off balance, which is why it is necessary not just to have lifelines attached but also to have proper attachment points in the cockpit well.

These must be sited to serve several purposes. First and foremost, as with any harness strongpoints, whether in the cockpit or on deck, the primary function is not to keep crew attached to the boat but to prevent them falling overboard in the first place, so they should be placed as near the centreline of the boat as possible. Secondly, it should be possible for crew coming on deck to be able to clip on before doing so. Thirdly, there should be a jackstay close enough to the cockpit to be able to clip on before leaving and, finally, there must be a suitable point for the helmsman to clip on, particularly if there is wheel steering and he is physically removed from the strongpoints at the for'ard end of the cockpit.

Peter Haward points out the dangers of using conventional carbine hooks with eye bolts as they can trip themselves. One solution to this is either to have the locking type of hook (with a crossbar or collar to prevent tripping) or to fit elongated eye plates where the span between the sides is greater than the width of the hook.

Deck security

Common sense apart, the four main lines of defence for deck security are grabrails, non-slip deck surface, safety harnesses and toerails and guardwires, in that order.

Grabrails are top of the list because, if properly sited and used, they should reduce the dependence on the others. They should be as high as practical to fall easily to hand, have more than enough room underneath for a hand or lashing, and should always be through-bolted or attached with machine screws tapped into holes with suitable metal inserts. Rails should run for the full length of the coachroof if practical.

Non-slip surface. The balance here is between effectiveness and convenience. But it isn't just the quality of the surface; it is also where it is distributed. Bill Anderson points out that you should always consider non-skid at an angle of 30 degrees. Surfaces which are not walked on when the boat is level suddenly become vital footholds – sloped coachroof sides, for example.

Most production boats have adequate non-skid surfaces, but where they fall down is on its distribution. Decks are normally well covered, except sometimes and

Sidescreens are vulnerable and they can rip or damage stanchions when full of water. They should be about 6in clear of the deck and be attached using shock cord to allow some give

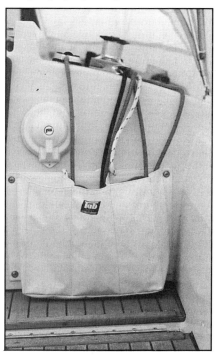

The trend for leading lines aft can result in some very muddled and chaotic cockpits unless stowage is properly arranged for falls and tails

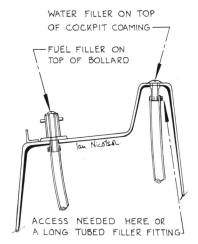
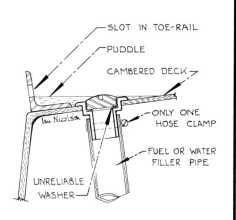

Tank filler caps are rarely completely watertight as the seals can perish, and often they are sited so as to be below the level of the nearest drain and therefore live much of their lives under water. Seals should be replaced every two years to prevent diesel or water contamination and the threads smeared with Vaseline. Alternatively (right), the filler can be on a coaming or even incorporated in a deck fitting such as a bollard

inexplicably a central strip on the foredeck, but the coachroof always takes second place. It is when working here, though, tying in a reef, or walking inside a shroud, for example, that deck crew are at their most vulnerable. The worst culprits are the front of the coachroof, often stylishly raked, the main hatch garage and deck hatches. The remedy is simple – strips of self-adhesive non-slip 3M tape available from most chandlers, spaced at less than a foot's width apart. Another potential hazard on the coachroof is when lines are led aft. Like jackstays, they can roll underfoot.

Bill Anderson highlights the dangers of a diesel spill completely destroying any non-skid properties and also the fact that any moulded non-slip surface will wear and need painting at some stage, though he considers that some modern non-slip paints are too abrasive. He recommends teak as the ideal compromise between non-skid and user friendliness, but 'unfortunately it doesn't come in 2-litre tins'. Peter Haward comments that non-skid properties clash with cleanability. Some diamond pattern surfaces only pay lip service to the non-skid requirement. 'Dirt will just slide off it, just like ropes, sail bags, winch handles and human beings.'

Harnesses on deck. The terminology, if properly used, is singularly appropriate for deck safety gear. The harness line may literally be your *lifeline* whereas the wires between stanchions are not lifelines but *guardwires*, guarding against, but not necessarily preventing, crew going over the side.

Although good grabrails and non-slip should prevent the lifeline being necessary, safety harnesses are the most important item of safety equipment to prevent man overboard. And as important as the harness and lifeline is the way they are used when moving around on deck.

The 1979 Fastnet Report showed that only 22 per cent of yachts had jackstays. Today the ORC Special Regulations (see extract page 58) specify jackstays for yachts competing in all categories of offshore race; and sadly, at the same time, it is rare to find them offered as standard equipment on production cruising yachts.

As stated in the ORC Regulations, jackstays should be arranged to allow access to all parts of the deck with the crew remaining clipped on. Anchorage points must be beyond question, through-bolted with backing plates, and wires should have proper swaged, spliced or Talurit terminals. Although stainless steel wire is widely used, if it is on the sidedeck it can create its own hazard. It will roll underfoot if trodden on, so should be sited either as close as possible to the coachroof sides or, if practical, along the top of the coachroof. For this reason, nylon webbing is becoming popular (especially, as it is quiet, with off watch crew below decks) but Bill Anderson recommends that webbing should be replaced every year as it is subject to ultra-violet degradation.

Plastic covered wires should be avoided, says Ian Nicolson, and they should terminate as far short of the forestay as the length of the lifeline allows – say 4ft with a 5ft lifeline – to keep the working area clear.

Toerails and guardwires. The toerail is a precise definition of its function. On a heeled deck it is there to provide a toehold. Some boats have none around the foredeck; some just pay lip service with a low, moulded rail. In both cases, the only practical solution is a third guardwire a couple of inches above the deck. The ORC gives a minimum height of 1in for toerails but the higher the better,

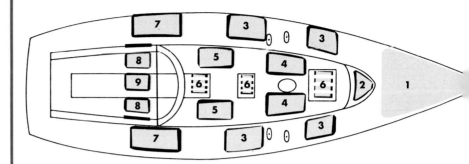

Vital non-skid areas on deck

Although the whole working area of the deck should have non-slip surface, some parts are particularly important. **1**, foredeck; **2**, front end of the coachroof, particularly if steeply raked; **3**, side decks either side of the shrouds; **4**, coachroof by the mast both for the up and over step when avoiding the shrouds and when working at the mast; **5**, either side of the coachroof for when working on the boom, reefing or stowing sails; **6**, on slippery deck hatches and garages; **7**, either side of the cockpit for the step in or out; **8**, cockpit seats for the step to or from the sidedeck; **9**, cockpit sole for the step on to the seats or going below

Two types of safety carbine hooks that cannot be accidentally tripped

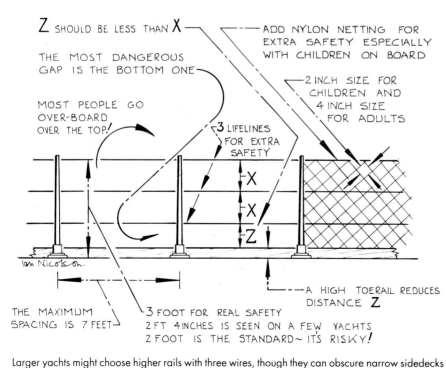

Larger yachts might choose higher rails with three wires, though they can obscure narrow sidedecks when the boat is heeled. However, stanchions higher than 28in will have to be tailor-made. On any boat netting is a good idea if small children are on board

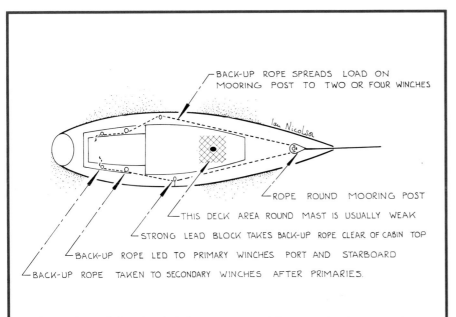

Modern yachts rarely have foredeck cleats strong enough for exposed anchorages and moorings or for towing. A back-up rope will help spread the load and utilise the strong point of the cockpit winches. This should be passed both round the cleat and through the eye of the mooring/tow, preferably with some chafe protection

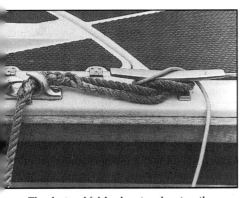

The cleat and fairlead are too close together. Imagine trying to take a line off the cleat when under load

A mast box will help keep the halyards in place and the deck clear

If the liferaft is stowed on chocks on deck, ideally the straps should go not to the chocks like this but to separate strong points. That way, if the chocks should fail when swept by a sea, the liferaft is still attached

especially if they graduate to low bulwarks with stanchions mounted on top giving extra height and security.

The ORC requirements for stanchions and guardwires, quoted earlier, are as good a standard as any for cruising boats. Ideally, they would be very much higher than the standard 24in but there is a problem of scale. As soon as the boat is heeled a high rail may make it impossible to walk along the weather deck.

Bill Anderson says that guardwires are 'a psychological rather than a real aid to security; most of the time they simply mark the edge of the playing area. The only time they are a real help is when someone is kneeling or sitting on deck'.

Ian Nicolson recommends Simpson Lawrence 28in stanchions if there are young, old, tired or seasick crew on board; ideally a third wire should be fitted, though deck width may be the restricting factor. He suggests that they should be mounted on wooden bases to give some spring and reduce deck cracking at the foot, and goes on to state that 50 per cent of yachts surveyed have below decks pads that are too small. He says that guardwires should be replaced every eight years and that plastic covered wires should be replaced as soon as any chafe or cracking appears in the sheathing.

The attachment of wires to pulpits and pushpits is important. If insulators are fitted they must be of the type that has a metal to metal link inside the insulation material. Perhaps the simplest arrangement is to shackle them on at the pulpit and have lanyards at the after end for insulation and tensioning. These have the added advantage that they can be cut when recovering a man in the water.

Deck hardware

It is unusual for production yachts to be fitted with mooring cleats that are adequate in number, size and strength for mooring or anchoring in exposed waters. The appropriately-named samson post all but disappeared with the advent of GRP and without major surgery it is not possible to adapt the modern yacht.

What can be done, however, when ordering a new boat is to specify oversized bow cleats with additional below decks reinforcement. And similarly, if you are likely to be cruising areas with exposed anchorages, it is not difficult to increase cleat size and under deck pads. In the accompanying illustration, Ian Nicolson shows how a deck cleat can be backed with lines to sheet winches for heavy weather anchoring or when being towed, and makes the point that the traditional method of towing with a line round the mast is fraught with danger with deck stepped masts. He also highlights the danger of lying anchored to a windlass alone as castings can fracture.

As the anchor is the last line of defence, anchoring arrangements must be as good as the anchor itself. This means a clear lead from bow roller to cleat, two substantial rollers, one for chain, one for rope with high cheeks and a means of closing them off with the cable in position.

Conclusions

As with so many other areas of the cruising yacht, there is little that can be changed of the physical structure of decks and cockpits, but it is the detail that can make all the difference between a hazardous and a comfortable and safe working area – such things as a dependable deck surface, easily used and strong harness points and deck gear that is up to the job. As the crew is the most important element in the seaworthiness equation, looking after him in terms of personal security and the level of efficiency at which he can operate will directly affect the safety of a yacht at sea. If he feels secure and comfortable then he is better able to make the correct decisions about the running of the boat.

Chapter 6

Into the interior

A vital ingredient in the seaworthiness of a yacht is the efficiency of her crew. And this is directly related to the layout and design of her interior

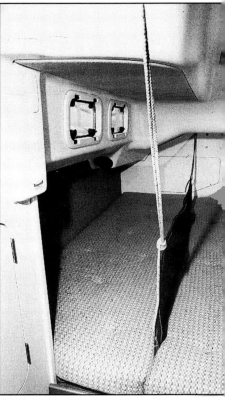

This simple modification turns a useless aftercabin into one or, at a pinch, two seaberths

TODAY'S YACHTS have a spaciousness and performance, length for length, which yesterday's designs could only dream about, yet many modern yachts are fitted out more as marine caravans or floating country cottages. Their owners may not have the slightest interest in crossing oceans or even contemplating overnight passages. For 90 per cent of the time there is nothing wrong with that, providing everyone is clear that a marine caravan is just what it says it is and that no one is under a delusion that it is a serious cruising yacht.

But our contention is that any yacht which, by its general size, commercial promotion or appearance might be considered for cruising offshore, should be designed and equipped to undertake this role properly.

Yachts cannot be put into compartments in the way that perhaps cars can. The analogy between the average family saloon and the Range Rover can only be taken so far. Both are designed for the open road, but whereas the saloon may cross a field in the right conditions, the Range Rover has special design features that more or less guarantees that it will cope with the most rugged terrain. Unfortunately, at sea the open road can turn into the roughest countryside almost without warning.

Equally unfortunately, good offshore accommodation is becoming increasingly incompatible with the commercial pressure to produce yachts which are comfortable in harbour and saleable at boat shows. As Bill Anderson ruefully recalled, of the four most comfortable seagoing interiors he had sailed in, none was designed later than 1970 and all were austere and uncomfortable for harbour use.

Sleeping

Comfortable, secure bunks are not a luxury, they are a necessity. Our panel was in general agreement that 20in was a good width for an offshore berth, with 24in being sufficient for harbour use. People are generally taller than they were, and a bunk length of 6ft 4in should, perhaps, be considered a target figure. While most builders get the width about right, few new yachts have bunks longer than 6ft 2in.

Bunks must be easy to get in and out of but, once in the sleeping position, a person must be expected to stay there. Canvas

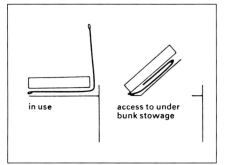

Design for a better leecloth. By attaching the cloth to the outboard edge of the bunk access to underbunk lockers is improved and the whole berth assumes a more rounded and comfortable shape when heeled

leecloths solve the problem of falling out and should be a standard fitting on every boat. They are neither expensive nor difficult to fit. They should be deep enough and wide enough to feel secure even when the boat is well heeled.

Our panel had quite a lot to say on the subject of leecloths. Ian Nicolson observed that they should be long enough to support the sleeper at shoulder, hip and knee, while leaving sufficient space at the head for ventilation. Peter Haward added that the cloth should be fixed so that it sloped inward towards the bunk in order to be completely secure. There was general agreement that cloths should be between 7½in and 10in high, with 10in air gaps at head and foot. Ian Nicolson recommended that leecloths and their securing lines should be designed to withstand a weight of 1,500 lb to allow for people being thrown about in the cabin.

The usual seaberths in a medium-sized cruiser are the saloon settees, but there are a variety of alternatives. Older yachts featured quarterberths, and these made excellent billets while on passage. Because the

1979 Fastnet Race Inquiry Report

This is an outline of the findings of the official report based on questionnaires return by 235 yachts

The compilers of the Fastnet Report were mainly concerned with features of interior design and constru which had a direct bearing on the safety and well-being of the crew. Simple matters of crew comfort a convenience were not considered.

Surveyor Ian Nicolson also emphasised that the report was based on a questionnaire which was fil and returned within a few months of the disaster. Many of the yachts were subsequently surveyed whe more internal damage came to light that was originally reported. However, the basic conclusions of th report remain valid.

Two major problem areas were highlighted. First, that very few boats had sufficient grabrails. Seco that yachts whose stowage, at all normal angles of heel was perfectly secure, found all sorts of things f tins to cookers flying round the interior when the yacht was inverted.

One yacht in five found that when hatches were closed against the storm, ventilation was inadequa Surprisingly for a fleet of offshore yachts, almost 10 per cent had insufficient bunks with leecloths/boa for half the crew.

Of yachts which reported damage to the interior, 29 per cent said it was caused by weakness of materials, 39 per cent said it was due to not being properly fixed.

Batteries and cookers were the chief culprits when it came to things flying around. But there were a reports of tins, which under normal circumstances never moved, becoming lethal missiles.

However, in total few skippers reported problems with stowage yet a majority said they spent a lot time clearing up the shambles after the storm, indicating that they regarded stowage failures as an occupational hazard. It is reasonable to speculate that, since a tidy ship will be a more efficient ship, t outlook is mistaken.

The report concludes: 'It is essential that heavy items of equipment should be locked in position by positive fastenings and should not rely on gravity to keep them in place.'

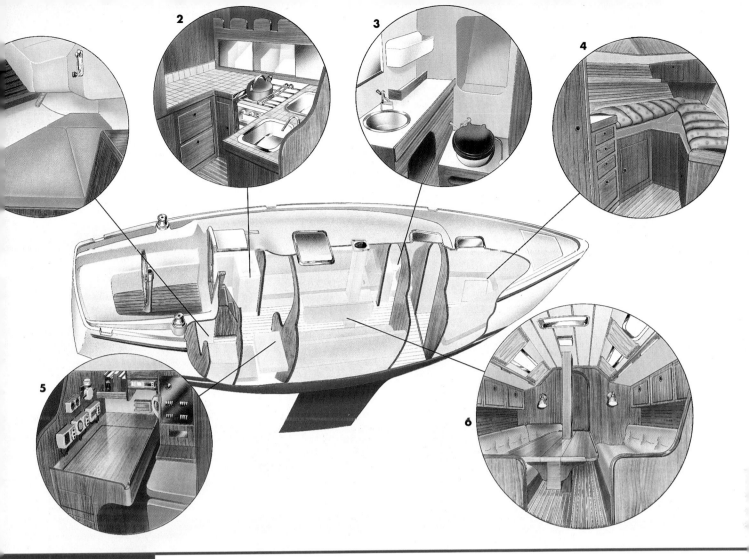

The living quarters

1. **The aftercabin** – the best part of the boat for sleeping offshore but often the worst designed. Split bunk cushions and a leecloth may be all that is needed

2. **The galley** – U-shaped galleys with plenty of secure stowage, a well gimballed cooker and well fiddled work surfaces can be used even in poor conditions

3. **The heads** – room to move when dressed in oilskins, rounded corners, plenty of grab handles and a wc accessible to use and clean/service

4. **Forecabin** – not a good sea cabin but, given space and good stowage, can make a good owner's harbour cabin. Front access to under bunk lockers is important

5. **The chart table** – room for a half-Admiralty chart, bulkhead space for instruments and a bookcase deep enough to take tall pilot books are the main requirements

6. **The saloon** – straight settees aligned with the centreline make ideal seaberths. Side lockers should have secure catches. Seat backs can be adapted to make pilot berths

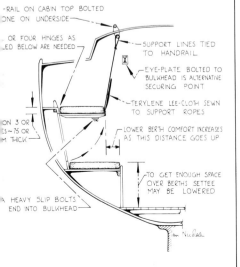

Seat back to pilot berth. Ian Nicolson's design for doubling usable seaberths

head of the bunk was often the chart table seat, the bunk was the preserve of the skipper/navigator.

The modern equivalent of the quarter-berth, the aftercabin, does not, as a rule, offer the same opportunities for rest at sea. Despite this, Bill Anderson believes it is potentially the best place for this purpose – motion is less, the sleeper is out of the way and it is quiet (unless the engine is running). The answer is to split the double cushion down the middle and install a leecloth so that one side, usually the outboard side, forms a narrow, secure pit. Even so, the curved, sloped nature of the hull in this area, caused by the low height of the berths, often makes sleeping uncomfortable on this side, while inboard you are forced into the airless, and often low, space under the cockpit right beside the engine.

An easy way to provide extra sea berths is to install pipe cots. At their simplest they can be two lengths of aluminium pipe with canvas stretched between them. The pipes are supported in cups on the fore and aft saloon bulkheads. For greater sophistication make a series of cups for the inboard pole to allow for different angles of heel, while for ultimate comfort make the height of the inboard edge infinitely adjustable by means of a block and tackle to the deckhead.

Pilot berths often make the most comfortable offshore sleeping spot. The disadvantage is that in smaller boats they are invariably installed in place of valuable stowage. On many boats, though, the settee backs, if they are not already arranged to do so, can be reinforced and given suitable support so that they can be raised and used as pilot berths.

Deep upholstery is not important for a good night's sleep at sea. Indeed, a pipe cot with no mattress at all can be quite comfortable. But it is generally considered that 4in, firm foam cushions are about right.

Owners with many offshore miles to their credit recommend that berths used regularly at sea should be covered in waterproof vinyl rather than fancy fabrics. This would certainly apply to quarterberths and also to dedicated pilot berths. Dual purpose sleeping areas, such as saloon settees, can be provided with an offshore set of removable, loose vinyl covers.

The galley
Design: The shape of the galley is crucial to its efficiency offshore and safety is of paramount importance. The cook must be secure. It is cheering to see that a good many yachts are fitted with crash bars in front of the cooker and that galley straps are also fairly common.

If the galley runs down one side of the boat, there is little to support the cook's back except a galley strap. He will need to use one hand to hold on with. L-shaped galleys, with the cooker in the middle of the long side, again leave his back unsupported. U-shaped galleys are ideal, providing good support and bringing all parts of them within easy reach.

There should be sufficient working space so as not to have to use the chart table when preparing a simple meal. The cook should be able to reach into all lockers and cupboards and use the cooker whilst standing in one place and being firmly supported.

Many galleys have sliding fronts for their lockers. These must be capable of being secured. Our panel commented that this sort of door was never hurricane proof and seldom even gale proof, particularly if the door ran athwartships.

Lockerage round the galley should be sufficient to hold food for two days at sea without needing replenishment from more deeply buried stores. Large open lockers are a liability and should be broken up into smaller areas. If this is impossible, items should be stowed in suitable plastic containers. Stacking baskets, such as breadbaskets, make excellent containers for bringing food on board and also for storing it and other bits and pieces of hardware.

When planning food stowage, try to devise separate areas for a hard dry store (coffee jars, for example), soft dry store (packet foods) and wet store (soft milk cartons, fruit and veg). Have an emergency store of long-lasting food (pasta, dried peas, etc) kept permanently on board.

Cookers: Most boats are fitted with LPG gas cookers these days. They are easy to use and economical to run. The proper copper piping and flexi hose, with gas shut-off valve accessible from the galley, should be fitted to all new yachts, but it would be something to check on a secondhand one.

Check, too, that the cooker can swing to its full extent without coming up short against the pipework and that the hose has not become dangerously chafed.

Ian Nicolson observed that 95 per cent of the boats he surveyed had badly clipped gas pipework and 4 per cent had no clipping at all.

All cookers must be gimballed in such a way that they are free to swing through 40 degrees as a minimum and ideally up to 70 degrees without hitting the hull or galley furniture.

The cooker must be firmly secured to its gimbals and be capable of being locked in the upright position. It is quite surprising to see how many ranges could, in a nasty sea, quite easily jump out of their fittings despite the findings of the Fastnet Report and the ORC regulations.

Sinks: In most cases, the sinks are too wide and not deep enough. If they are too big they will use up gallons of water every time you want to wash up a plate and cup. A basin a foot square and 8in deep is about right. It is useful to have a sea water tap so that fresh water can be saved. A narrow drainage sink is valuable is space permits.

Working space: Worktops should be 36in off the sole and can be made safer by covering them with non-slip material. This can be bought in a roll and cut to fit. In the absence of this, though, damp tea towels are effective in normal weather.

Cool boxes: Large cool boxes are nice to have but can get into a terrible muddle if not divided up by shelves and partitions. Even if fitted with fiddles or drainage channels round the top, cool boxes collect liquid at the bottom from melted ice, spilt milk and other sources. This must be drained out but not into the bilge where it collects and smells. A good arrangement is a pipe with a double U-bend from the drain plug to a bottle which can be emptied regularly. This keeps the cold air in but allows liquid to drain out.

Gash: There should be somewhere for the bag of rubbish to be stowed where it cannot spill out over the floor, either in a convenient locker, or an external fitting that is not going to be knocked out of place by someone pushing past in a hurry.

The heads
Perhaps the most sensitive area of ship board life is the heads. A compartment of the right size and properly designed for use by crew fully kitted out in oilskins can transform life on a family yacht.

Positioning of the bowl in many modern yachts calls for great athleticism from the crew. Complaints from our panel included bowls too close to the door, under sidedecks and at awkward angles. It must be accessible not just for all the uses to which it will be put, but also for cleaning and servicing.

The essential features of a loo compartment are: room to move in oilskins; the ability to wedge yourself in against the motion of the yacht; grab handles high and low; a total absence of sharp corners and projections (particularly at head height); accessible seacocks; a bolt or catch to the hold the seat and lid up, and a waterproof loo roll holder.

These considerations are, on the whole, more important than whether the heads is in the currently fashionable position aft or in the traditional situation between the saloon and forecabin. The former is often very effective and has the advantage of greater headroom and less motion when at sea, provided sufficient overall space is allocated to it.

Stowage
One of the lessons learnt from the Fastnet Inquiry was that careless and inadequate stowage caused a number of severe head injuries when heavy articles, such as tins of food, were flung from lockers.

Knockdowns are, happily, rare occurrences, but the fact remains that they can and do happen. In those situations only the items that are securely battened down will remain in place. Heavy items, such as batteries, water and fuel tanks, anchors, tool boxes and so on, must be secured individually and should be kept as low in the boat as possible. Small, loose items should be securely restrained.

Fuel and water: Tanks for these should be positioned for best effect in terms of trim and ballast. Fuel should never be stowed within the living compartment. Usually, the tank is found under the cockpit sole or in a cockpit locker close to the centreline. Provided it is properly secured, this is probably as good as you will get.

The ideal place for a water tank is in the bilge over, or astern of, the ballast keel. In modern yachts with shallow bilges, there is a tendency for tanks to be situated under the saloon seats. From the point of view of trim and ballast, a tank either side in the saloon is not too bad, but a single tank on one side, or a tank forward under the forepeak bunk, can

---

ORC Special Regulations (category 2)

This is a précis of the requirements for interior stowage and security contained in these regulations. Copies of the regulations are available from the RORC, 19 St James's Place, London SW1A 1NN

5.2 Inboard engine installation shall be such that the engine, when running, can be securely covered, and that the exhaust and fuel supply systems are securely installed and adequately protected from the effects of heavy weather. When an electric starter is the only provision for starting the engine, a separate battery shall be carried, the primary purpose of which is to start the engine.

5.4 Ballast and heavy equipment. All heavy items, including ballast and internal fittings (such as batteries, stoves, gas bottles, tanks, engines, outboard motors, etc) and anchors and chain shall be securely fastened so as to remain in position should the yacht be capsized though 180 degrees.

7.11 Toilet, securely installed.

7.2 Bunks, securely installed.

7.3 Cooking stove, securely installed against a capsize with safe, accessible fuel shut off control capable of being safely operated in a seaway.

7.53 At least one securely installed water tank discharging through a pump.

8.1 Fire extinguishers, at least two, readily accessible in suitable and different parts of the boat.

8.21.1 Bilge pumps, at least two, manually operated, securely fitted, one operable above, the other below decks. Each pump shall be operable with all cockpit seats, hatches and companionways shut.

A snug and secure interior: straight settees for effective seaberths, a solid table and stout pillars to hold on to

be a disaster. A full water tank may well be the heaviest single item onboard a boat other than the keel and engine. To have this weight in the bows is bound to increase pitching and upset trim. Owner's demand for fresh water is increasing with many yachts equipped with showers, hot and cold pressure water and so on, but its provision must be kept in proportion with the need to stow other items of gear and equipment.

Water tanks of more than 20 gallons capacity should be fitted with baffles, particularly those under bunks. If not, the slop of water under the ear is enough to keep the deepest sleeper awake.

Batteries: These must be securely fixed. Battery acid causes severe burns and the gas fumes are highly toxic. Strong, removable battens across the top or webbing straps are a simple way of achieving this, though they are best stowed in special, vented boxes with stout lids and waterproof bases.

Under bunk stowage: Even with tanks under the berths, there is usually space left over here for general stowage, and this is the best place for heavy items such as tool boxes, spare food tins, etc where their weight is contributing to stability. Often this area is used for clothes, but this is not a good idea since a modern shallow bilged yacht often gathers water in these lockers. Clothes are best kept in lockers behind the settee backs or under the sidedecks.

Best access to under bunk stowage is via hatches in the bunk fronts. This avoids disturbing crew sitting or lying on the bunk. But there should be top access, as well, for putting in large items and reaching things at the back. Large access hatches are better than small ones, and it is a good idea to make catches for them so that they remain in place in the event of a knockdown.

On the subject of personal stowage, Bill Anderson observed that lockers for this purpose should be small and plentiful to avoid the midshipman's chest syndrome – everything on top and nothing accessible.

General stowage: Lockers in yachts of all ages suffer from the sin of failing to keep their contents in place. The lockers behind the saloon seat backs may take the form of three holes covered only by the seat back attached by Velcro. The moment the ship heels and bounces from wave to wave, the cushions fall off and the contents spill over the saloon sole. This is not in itself dangerous, but precious dry clothing is liable to get wet, and it adds to the general chaos if you have to trample through piles of underwear to get on watch.

Waist-high lockers in the body of the saloon are not the ideal place for stowing tinned food, as the sheer force of tins on the move can push open the door or trigger the finger catch from inside. All locker doors should have positive fastening and not rely on magnetic, Velcro or other push-pull catches. If this sort of locker is the only stowage place for tins, etc, then fit those cup-type drawer pulls upside-down inside the locker fronts so that they protect the catches from a tin triggering them. All locker shelves should be fitted with fiddles to restrain the contents independently of the door.

Peter Haward recognises that tins will be stowed in this sort of locker, but strongly recommends that they be removed to a lower and more secure place at the approach of really bad weather.

If tins are to be stowed in the bilges, remember that labels will get wet and soak off. Remove the labels deliberately and mark the tins with an indelible felt-tipped pen.

Grabrails and fiddles

The old sailor's maxim runs 'one hand for yourself and one for the ship', but this presupposes that strong, properly secured grabrails and handles are available in the right places throughout the interior.

The absence of good hand-holds in the saloon or galley may not be apparent until they are needed. You should be able to walk the length of the saloon holding on to something firm at all times and at all angles of heel. Deep gullies or rails running along under the side lights are in the best position when a boat is well heeled. Rails fixed to the deck head should be bolted through to the grabrails on deck.

In addition to these, grab pillars at the chart table and galley are invaluable in the part of the boat where people are most often standing when at sea. Equally, the companionway steps should be provided with handles to aid entry and egress to the saloon.

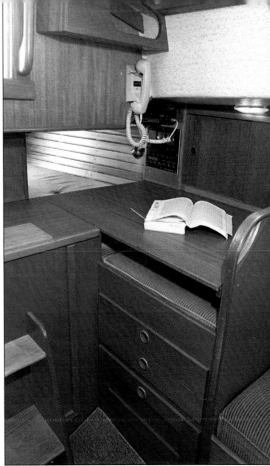

On a small boat a stand-up chart table across the head of a quarterberth is a practical solution to navigating in the Decca age

It is often the case that, having provided a minimum of handles and rails in the saloon area, the builder considers his job done. This is not the case. The heads, aftercabin and forecabin, of which the former is the most important, should all be given careful consideration.

Many yachts have nicely sculpted fiddles which act as perfect ski-jumps for plates, pans of boiling water and glasses, propelling them with added vigour into the laps of hapless crew.

Fiddles should have vertical faces on the inward side and be at least 3in high. Ian Nicolson recalled that, during his apprenticeship, rails 2in high were considered sufficient for the heavy, steady yachts of those days (1945). Today, he says this should be increased to as high as 4in for deep sea use in modern, lively boats.

Fiddles around the sink, galley and general working surfaces must be deep enough and strong enough to act as grab bars. They should never enclose a space completely, and gaps should be left at the corners for cleaning purposes.

Removable fiddles are sometimes found, particularly on saloon tables, the idea being that they are not needed in harbour and the table is consequently more comfortable and convenient to use. When new they work well but the pins and holes inevitably work loose, even if brass lined, and give way at the vital moment.

Saloon table

Bill Anderson argues that on boats under 29ft, a saloon table is an unnecessary luxury. It takes up most of the saloon, making the interior difficult to move around while the oncoming watch can find nowhere to stand

and put on their trousers. At sea most people eat off their knees anyway, and few tables are strong enough to withstand a man falling against them.

Many owners, particularly those with families, would probably reply that a table was too important in harbour to be discarded completely. The solution is a demountable one.

When in position any table must be firmly bolted to the floor and its flaps capable of being securely held down. Velcro is not sufficient for this. Bolts, cabin hooks of the non-rattling type, or other, reliable mechanical locks are the answer.

Chart tables

The subject of the orientation and design of chart tables will cause heated debate in any yacht club bar.

In our view, a good table is aligned so that the navigator is facing forward and has room to spread out an Admiralty chart folded once. It will have shelving for almanacs and pilot books, sufficient bulkhead space for electronic instruments, and stowage for charts and navigating instruments. Most owners also like to have room in the chart table area for much of the paraphernalia of running and maintaining the boat, from binoculars to repair kits, small spares (shackles, spark-plugs, etc) and manuals.

However, the facts of life are that owners very often have to make do with what the builder has managed to fit in. Athwartships, and even aft facing, tables, though not ideal, are usable. The athwartships table can be particularly trying. On one tack the working surface will be tilted well away from you and on the other gravity will cause you and the chart to end up in the galley bilges. However, in these Decca-assisted days, some owners are finding an athwartships table, with well fiddled edges all round, at which you stand, not sit, and under which can be provided all the stowage mentioned above, is a practical solution. Chart work is much simplified by the Decca, while detailed passage planning will take place on the saloon table before the voyage begins.

There should be something to hold dividers and pencil. Those fancy pencil holders get broken within a day, but a cheap way of fixing your own is to screw a small length of drain pipe to the bulkhead. Plotters or parallel rulers can be held in place by two wedges screwed on the bulkhead. Blobs of Blu-Tak dotted round the chart table make excellent temporary holding points for dividers, pencils and rulers pressed on to them.

All electronics are susceptible to damp and many boxes are at the best only splashproof. These can easily be protected from spray and the navigator's wet hands by a simple covering of plastic sheeting, through which the buttons can still be operated. The switch panel can also be protected from damp and inadvertent switching on by the navigator's elbow using a hinged sheet of rigid, clear acrylic.

The smaller the boat, the harder it is to fit a permanent navigation area. Bill Anderson, predictably, believes that it is a waste of space to fit one to yachts under 25ft. Instead, he recommends a board on a bunk.

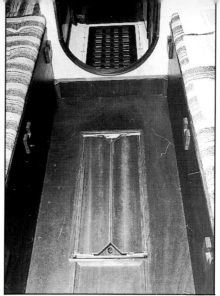

This ingenious folding system solves the problem of tables taking up too much room in saloons

The companionway

The companionway steps are an area of great potential danger. Ladders held in place only by a couple of piston bolts into the sole are not very secure, as they will weaken in time and eventually break. U-shaped chocks on the sole, in which the feet sit, will spread the load and prevent the steps slipping if the bolt is either not properly shot home or breaks.

The treads themselves should be deep and wide enough to take a large man's booted foot safely. Steps with sides are preferable for three reasons. First, the sides are, in themselves, integral hand-holds. Second, they act as guides and restraints for feet when coming below. Third, open-sided steps have sharp edges against which to fall.

The sole

Glossy, highly polished floors belong in ballrooms not yachts! They may look attractive, but water, let alone anything else, turns the area into an ice rink.

Matt polyurethane varnish, which looks attractive, is adequate under most conditions, but a sprinkling of sand or a proprietary non-slip surface, particularly at the foot of the companionway, is better. Treadmaster is completely skid-proof, but it chews up your clothing in next to no time. Strips of stick-on 3M non-skid material is cheaper and equally effective but needs replacing every couple of years. Carpets are only suitable for use in harbour.

The question of whether bilge access panels should be screwed down to prevent them from flying about in the event of a knockdown or left loose to be easily removed for emergency access to strum boxes, keel bolts and bottles of wine, is another subject for debate. Our panel elected for a couple of small inspection panels and the rest secured.

Lighting

Efficient lighting is important during night passages. The navigator should have an adjustable chart light, arguably with a red filter and/or dimmer switch, or an Aquasignal-type polarising light. Overhead lights should have dimmers. The key thing, though, is to have just enough light for those coming on watch to see what they are doing without it shining directly into the eyes of those already on watch.

Ventilation

Important areas to ventilate include the heads, galley and aftercabin. Ventilators should be fitted with dorade or baffle boxes. One or more solar-powered electric vents keep a boat sweet even when it is shut up for a long time.

As Bill Anderson observed, there is a trend towards using small hatches or skylights in areas such as the heads instead of conventional ventilators. A ventilator is a device which allows the free passage of air and discourages, or at best prevents, the passage of water. A hatch is not a substitute for a ventilator. An offshore yacht must be capable of being kept properly ventilated, even with the cooker on, with the main hatch and any other deck hatch or port closed.

A good place to put a ventilator on a modern yacht with poor air circulation is in the anchor well. This will draw air into the forecabin and is extremely unlikely to allow water through, even in severe conditions, provided the well is properly drained. Prudence dictates, though, that it must be capable of being securely closed with a screw on the inspection lid

Noise

Extraneous noise, particularly unidentifiable rhythmic ones, is one of the most common causes of insomnia. One of the commonest sources of this are tins and bottles rattling in lockers and other small items moving in their stowage. Keep a selection of rags, sponges and old socks to deaden the sound. Lengths of shock cord round the sides and backs of lockers act as noise fenders.

Right: whether traditional or modern the interior of a yacht must allow the crew to work and rest at sea and be comfortable in harbour. With some modern designs the latter compromises seagoing comfort and security

Chapter 7

Engines and engine reliability

The inboard engine has become central to the running of the modern offshore yacht. Apart from providing motive power and electricity for the lighting and electronic systems, it also plays a key role in the seaworthiness of the yacht

THE RELIABILITY of the modern marine diesel engine means that it has become a major part of the overall seaworthiness equation. In difficult conditions it can be the one factor that allows a yacht to get clear of trouble, the lee shore for example, and avoid tricky or dangerous situations in the first place. It also forms part of the armoury the yachtsman has at his disposal for use in heavy weather.

However, like any item of equipment which a yacht depends on, it will only start and continue to function if properly fed, watered and maintained. Poor conditions will reveal any fault quickly and right at the time we are likely to need the engine most. The engine will be under greater strain, there will be more water and damp around and fuel and oil will be stirred up, increasing the danger of dirt and water entering the fuel system.

Yachtsmen traditionally have a serious mistrust of marine yacht auxiliaries which is less justified today with the modern lightweight marine diesel. This mistrust stems from the days when the auxiliary was just that, a tiny petrol-fuelled beast lurking in the bilge and providing just enough power to enter harbour and berth. The fact it was there meant that it was expected to work and, more often than not, when committed to a manoeuvre that required engine power it would sulk and refuse to start or, worse still, if running would putter quietly to a halt.

Petrol engines and a salt-laden, damp atmosphere have never enjoyed a happy relationship. Apart from problems with carburettors and fuel, their electrical systems, whether magnetos, coils or spark plugs, are particularly prone to failure, and yet it was not until the early 1970s that the diesel auxiliary became the norm. Today, the reliability of the petrol engine, especially outboards, has increased enormously, yet still the yachtsman sailing with a petrol auxiliary is likely to motor defensively, with one hand ready to hoist sail or to anchor.

The modern diesel engine, though, is a totally different matter as it relies to a far

It is in heavy going that engines are put under greatest strain and when any faults, from loose connections to fuel contamination, will make themselves felt the most

Troubleshooting

1. Exhaust checks: is the cooling water flowing (exhausts through the transom are much easier to check)? Is there any smoke? Black smoke means overloading, lack of air (check air inlet/filter) or worn injectors; blue smoke means oil is being burned (worn engine); white smoke indicates that water is finding its way into a cylinder.

A regular check should be made on the exhaust hose, from engine to outlet, for chafing. Ian Nicolson reports that one in every five boats he surveys has evidence of this. Inspect the hose where it passes through bulkheads and look for any sharp objects that might press on the exhaust. All connections (this is true of all engine water piping) must be double clipped as each one is a potential leak. The results of a broken or leaking exhaust pipe are extremely messy and, if undetected, potentially dangerous because of the fumes.

Even with a swan neck in the exhaust, it is possible for water to siphon back into the engine in heavy weather with following seas so some kind of blocking arrangement for the exhaust (perhaps simply a wooden plug padded with rags) should be arranged

2. Contaminated fuel causes the vast majority of problems with diesel engines. You should do everything possible to ensure a clean supply right from the pump through to the injectors.

Whatever the source of your fuel, it should be filtered as it goes into the tank using a funnel with a fine mesh filter or, failing that, a funnel and a piece of stocking nylon. The filler itself should be sited so as to be clear of deck water, be well separated from the water filler and very clearly marked. The tank breather is another possible water inlet and so should be well protected, preferably in a position where any overspill will not go on to the decks. Diesel on the decks is very difficult to remove and makes the surfaces extremely slippery and dangerous. Before filling with diesel a good ploy, especially with wooden decks, is to empty a bucket of water around the filler and breather so that any spillage 'floats' off.

Some tanks have a sump **(13)** built as a sediment and water trap and with a drain plug at the lowest point. Better still is a drain tap so that part of the routine checking of the engine will include draining off a cupful of fuel to remove this and check for impurities.

A baffle plate or plates in the tank will help prevent surge and sediment being stirred up.

Running out of fuel is a surprisingly common occurrence, mainly because many tanks have no kind of fuel gauge. A sight glass is one solution but a good discipline in any event is to log engine hours carefully between refills. With tanks that are nearly empty, sediment will be more of a problem, so it is good practice to refill when less than a quarter full – and always carry a spare two-gallon container of fuel.

There are two schools of thought on laying up tanks. The one says that a tank left full will cut down condensation and reduce corrosion, whilst the other argues that, over the winter, bacterial and other growth can form in and contaminate the fuel and that static fuel in supply lines and injectors can itself cause corrosion. Either way, there should be an inspection plate in the tank large enough to allow it to be cleaned and checked for internal corrosion

3. A fuel tap near the tank is essential so that it can be isolated in the event of fire or fuel line failure

4. The siting of the **glass bowl water/ sediment trap** is important. It should be easily accessible for regular checks with enough space below to allow water/sediment to be drained off easily

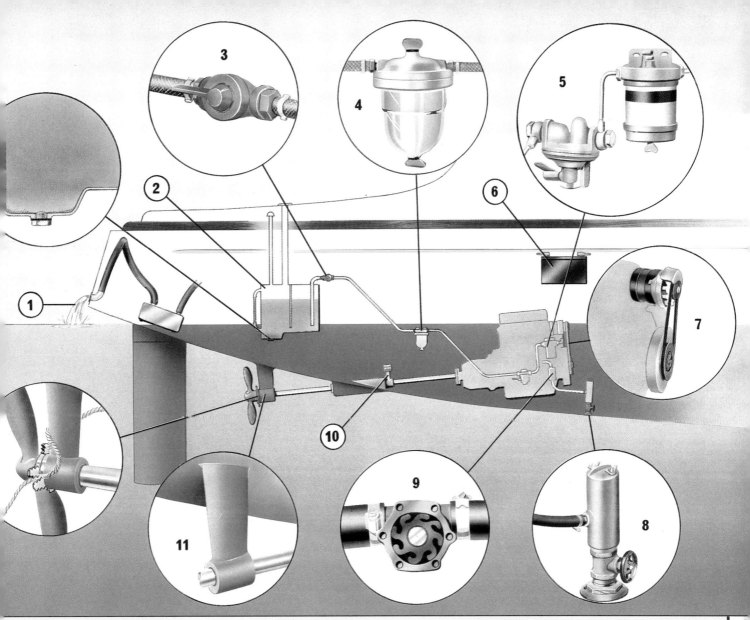

5. Lift pump and main filter: always carry at least one, preferably two spare filter elements. This filter is the main line of defence in preventing contamination reaching and blocking the injector pump. David Thomas recommends a twin filter system, with the filters in parallel and a diverter tap so that one can be left in line whilst the other is cleaned

6. Manual starting has never been high on the list of priorities when designing engine installations on sailing boats. Even on those engines fitted with hand start gear, there are few people who will be able to crank a diesel of more than 15hp and exceptionally few who can manage anything over 20hp. Because of this, it is better assume that starting will always be by battery and provision made for this. Ian Nicolson recommends buying 'the biggest batteries your bank overdraft will permit' and that they should be marine rather than automotive units. Even on small boats with few electrical demands, a second, dedicated engine starting battery is a good idea unless you are certain that you can hand start easily

7. Drive belts for the alternator and, in some cases, the water pump are all too easily neglected. At the first sign of wear, replace them and keep the old belt as a spare. Tension is important but not always easily achieved in the confines of an engine compartment. A long screwdriver or tapered length of wood used as a lever between engine and alternator usually works, first slackening off the alternator securing bolts and then tightening them when there is enough belt tension

8. A blocked water inlet won't stop an engine running but it will very quickly damage the pump impeller **(9)** and soon after major engine components. The first signs of a blocked water inlet will be a change in the engine note as the exhaust starts to run dry which can be confirmed by a visual check. The temperature alarm, if fitted, should sound before any damage is done, at which point the engine should be stopped immediately.

Whilst there is little apart from a vigilant look-out that can be done to avoid blockage from plastic sheeting, regular cleaning of the inlet strainer will prevent cumulative blockage from weed and minor debris.

Changing the water pump impeller is quick and easy and a spare should always be carried

10. There are many types of **stern gland** from the maintenance free through water lubricated to conventional packed gland. In most cases, there are two important elements, the gland itself and the greasing system. It is far better to tolerate a few drips from the stern tube than to over-tighten the gland, as this will cause shaft wear and in time far worse leaks. The first line of defence should be the stern tube greaser – better to over-grease a slightly loose stern gland than to under-lubricate one that is too tight. When motoring a half turn (until you feel pressure on the greaser) every couple of hours should suffice

11. On fin-keeled boats, the **P bracket** is the most commonly used shaft support which, whilst it gains in terms of simplicity suffers in that, mechanically, it is difficult to install well with adequate internal hull support. P bracket failures are not rare and, apart from leaving a hole in the boat, a broken bracket flailing around can cause considerable damage itself to hull and stern gear. Visual checks inside the hull – looking from cracks and any water seepage – will give early warning and every time the boat is dried out a thorough inspection should be carried out. Look for cracks, check for any movement and tap with a hammer to see if it sounds solid.

The cutlass bearing in the bracket can wear and if you are able to move the shaft appreciably it is time to have it replaced

12. The most reliable engine in the world is useless if the **propeller fouls**. The majority of propeller snarl-ups are self-inflicted in that they are caused by lines carelessly allowed to go over the side or mooring lines over run. Rigid checks, especially when motor-sailing, should be made and if you suspect you have just motored over a line take the engine out of gear immediately. Shaft cutters, as illustrated here, claim to help keep props free should they become fouled with light lines or fishing nets

lesser extent on electrical power, is far more tolerant of hostile conditions and uses a much safer fuel. Purists and traditionalists might consider this heresy but today the *reliable* auxiliary plays a major role in the seaworthiness of the cruising yacht.

The combination of motor and sail is a powerful one when, for example, trying to claw off a lee shore or when making to weather in strong winds. That lifelong small boat sailor, Humphrey Barton, mentioned 'the irresistible combination of sail and power'. Though well disposed towards their engines, some yachtsmen seem not to appreciate the value of using their sails and engine together. They tend to stow all canvas before starting up. That way they need to run at high revs, whereas sail assisted by power requires a surprisingly low running speed to achieve the same progress and will be easy on the fuel.

When making to windward, excellent results come from using the engine instead of the foresail. You will then point higher yet keep up a similar speed, still with remarkably small fuel consumption. In this way, the engine allows the less experienced to motor out of trouble, perhaps of their own making. It will assist in making port rapidly in deteriorating conditions and the noise of a running engine can have a remarkably calming effect on an anxious or frightened crew.

Some words of caution are worth mentioning on the question of sail and power to windward. Some smaller modern diesels may lose oil pressure when heeled beyond 25 degrees because the oil feed can suck air – different engines have different limits and it is worth checking your own engine manual for details. An oil pressure gauge will give early warning of this and many engines today have an audible alarm. The answer is to reduce sail to achieve a more reasonable angle of heel. Some larger engines can be specified with a deep sump to get over this problem. The same goes, to a lesser extent, for fuel tanks. A tank that is only partly full means that air, not to mention sediment, may get sucked into and stop the engine. If you know the weather is going to be bad, make sure your tanks are topped up. And no engine is going to carry on working if loose lines on deck go over the side and wrap themselves around the propeller.

It is the question of reliability and dependence on engines that is central to this argument. Unless you can be as near 100 per cent sure as possible that your engine will start on demand, whatever the conditions, and continue to function, then your confidence is dangerously misplaced; that dependence is likely to get you into more trouble than if you sailed engineless. And it is regular checks and maintenance which are the key to reliability.

Consider for a minute fishing boats that put to sea almost every day of the year under engine power alone. Most are single screw craft that rely wholly on their engines without the luxury of back-up in the form of sail. This total dependence means that maintenance and a thorough familiarity with the engine's working is a daily routine. The very fact that the engine is worked daily as opposed to occasional weekend outings also increases reliability – a diesel engine sitting idle for long periods will deteriorate more rapidly than one used frequently. Nor do diesels benefit from running at low speeds or idling for long periods. When battery charging, for example, it is better to run the engine in gear to put load on the engine and allow it to heat up quickly. Every diesel will benefit from being worked at full throttle from time to time.

On sailing yachts, the engine is seen as the secondary form of propulsion and this alone is likely to make it less reliable because we relegate it unconsciously to second place when it comes to understanding what makes it work and what is liable to go wrong. With sails, unless they are set properly the boat simply won't go or will perform badly and even the least experienced yachtsman has a rudimentary understanding of sail trim. With engines that understanding often finishes at the starter key.

A strange statistical anomaly with light aircraft engines illustrates the point. Aircraft engines have to be serviced regularly to rigorous standards and, however apparently healthy, rebuilt at a pre-set number of engine hours. The same standards apply to both single and twin-engined aircraft. Yet the number of failures in relation to flying hours

1979 Fastnet Race Inquiry Report

● Several yachts used their engines during the storm to help maintain steerage way, to keep the yacht at what was considered a safe angle to the waves or to improve pointing to make an offing from the Cornish coast. At least two dismasted yachts retired under power unaided. Of the three yachts that picked up survivors from other yachts or liferafts, two used their engines to improve manoeuvrability. Some competitors who tried to use engines to manoeuvre during the storm reported being unable to do so because they had no electrical power available for starting.

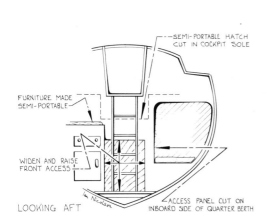

Getting at the machinery

Engines are often shoehorned into the tiniest and most inaccessible places and, whatever the efforts of the engine makers to put service points at the front of the engine, installations frequently leave much to be desired.

In bad cases, fairly major surgery may be needed by cutting a hole in the sole and fitting a watertight hatch. Side access to the engine may not normally be necessary but it is just as well to have side access panels in the quarterberth/aftercabin/galley/cockpit locker for the unusual or unexpected repair

ORC Special Regulations, Category 2

We have chosen these regulations, which are safety and equipment requirements for offshore racing yachts, as a model for this series. Category 2 is for yachts taking part in races 'of extended duration along or not far removed from the shoreline or in large unprotected bays or lakes where a high degree of self-sufficiency is required but with a reasonable probability that outside assistance could be called upon for aid in the event of serious emergency'.

Below is a précis of the requirements for engines and electrical equipment with the paragraph numbers referring to those in the Regulations. Copies of the booklet are available from the ORC, 19 St James's Place, London SW1A 1NN (price £2.00, plus 20p p&p).

● Inboard engine installations shall be such that the engine when running can be securely covered, and that the exhaust and fuel supply systems are securely installed and adequately protected from the effects of heavy weather. When an electric starter is the only provision for starting the engine, a separate battery shall be carried, the primary purpose of which is to start the engine (5.1).

● Each yacht fitted with a propulsion engine shall carry a minimum amount of fuel in a permanently installed fuel tank. This minimum amount of fuel shall be sufficient to be able to meet the charging requirements for the duration of the race and to motor for at least eight hours (5.1).

● All yachts shall have seacocks or valves on all through-hull openings below LWL except integral deck scuppers, shaft log, speed indicators, depth finders and the like; however a means of closing such openings, when necessary to do so, shall be provided (6.51).

● All yachts shall have shut-off valves on all fuel tanks (8.9).

This installation gives excellent front and side access but even though there is facility for hand starting the position is poor for a good swing

Making the propshaft safe

Every year a number of boats lose their propeller shafts through the stern tube when the coupling works loose. The result is a sudden deluge which is usually hard to reach and difficult to stop. The standard precaution is a hose clip (or two) clamped firmly round the propshaft close to the stern gland. Better is to prevent it in the first place by locking the coupling bolts and securing the shaft to the coupling with a taper pin or similar

is greater on an engine fitted to a twin than one fitted to a single-engined plane. Perhaps it is the total dependence on a single engine that makes engineers and pilots uncon sciously just that little bit more careful when servicing and checking. Although our dependence is less – boats are hardly likely to plunge to the bottom of the ocean when the auxiliary stops – the way we view our engines has a major bearing on reliability.

As with all aspects of seaworthiness and seamanship, the most important item of equipment on the boat rests between the ears of the skipper. Whilst common sense and a strong sense of survival come top of the list, knowledge comes a close second. An understanding of how an engine works, what is most likely to fail or cause problems and the ability, to rectify them plus a thorough familiarity with the layout of the engine and the battle is half won.

But before problems occur, the over-riding philosophy should be to prevent them happening in the first place. There will, of course, always be the unforeseen or unexpected but for most mechanical failures it is regular routine and preventive maintenance that is the best cure. Apart from regular servicing according to manufacturers' instructions, it is a good idea to carry out a pre-passage 'visual' in the same way as pilots carry out a pre-flight visual check on their aircraft. Does the fan belt look loose/frayed? How are the oil levels? Any obvious oil or water leaks? Is there water in the glass bowl filter? Is there enough water in the cooling system? Is there grease in the stern greaser? And so on.

Sizes and capacities

Whilst it will be of little consolation if you already sail a boat with a true auxiliary (ie power enough just for berthing or making a few knots in calms), as opposed to one that offers true motor-sailer 100/100 performance, as a general rule of thumb the minimum size of engine that will offer assistance in heavy going or allow the boat to achieve hull speed without full throttle in flat water is between 4 and 5bhp per ton of displacement, depending on hull form. This is a useful figure to remember when buying new or second-hand, especially bearing in mind that the under-powered boat might look a more attractive proposition in cash terms until you consider that to re-engine her will run into thousands of pounds. The under-powered boat can expect little help from her engine, however reliable, as an aid to seaworthiness.

Fuel capacities obviously depend on engine size and consumption, but the ORC Special Regulations should be taken as an absolute minimum (enough for eight hours' motoring). Apart from the inconvenience of having to refuel more often, a tank capacity that offers 24 hours or more of motoring at cruising revs is a sensible size. Remember too that consumption at full revs can be nearly double that at cruising speed (say threequarters maximum revs) and in heavy weather maximum power may be needed for prolonged periods. Diesel engine revs are governed and they don't suffer (in fact many thrive) from full throttle treatment unless injectors are badly set or calibrated, in which event the clouds of sooty black smoke from the exhaust will tell you when to throttle back. And as important as fuel capacity is the discipline of carrying a minimum of two gallons in a spare can, clearly marked to distinguish it from outboard and other fuels.

Engines and rescue – the telling statistics

RNLI rescue statistics and yacht auxiliaries
The extent to which yachtsmen rely on their engines is reflected in the 1988 RNLI rescue statistics for yachts with auxiliary engines. Whilst it is questionable whether *sailing* yachts with engine problems should be seeking assistance at all, the figures are illuminating.

There were 484 launches to sailing yachts with auxiliary engines requiring assistance. Of these a staggering 37 per cent were for engine related problems as below:

Machinery failure	129
Fouled propeller	37
Out of fuel	15

By comparison the figures for other categories were:

Stranding	101
Adverse conditions	53
Steering failure	36
Sick crewman	24
Sail failure or dismasting	22
Fire	12
Man overboard	6

Two members of our panel described these figures as 'amazing' and 'shocking', drawing one comment that because today's well-equipped yacht relies so heavily on mechanical and electrical systems, from pressure water to electronic navigators and roller reefing headsails, this very reliance appears to offer insulation from the environment and reduce self-sufficiency. And with less self-sufficiency, when there is a major equipment failure such as the engine the crew are less able to cope with the problem.

Some yachts fit 'day tanks' in addition to the main fuel tank which offers two main advantages. These tanks, perhaps just a few gallons in capacity and mounted high, are filled via a manual pump from the main tank. In the event of fuel pump failure they can be used to gravity feed the engine and they also allow a careful check to be kept of fuel stocks and consumption.

Learn, familiarise and practise

The average motorist today has much less idea of what happens beneath the bonnet of his car than 10 years ago. The reliability of the modern car means that he has far less reason to poke around the engine compartment and the complexity of today's car engine is likely to deter him in any event. This has had a knock-on effect, in that the lack of knowledge through necessity means that the yachtsman too is likely to have less knowledge about his boat's engine, especially as it is probably an unfamiliar diesel.

Learning about marine engines is the first step. Many engine manufacturers or importers run courses for boat owners, and these are extremely valuable and well worth the effort to attend. Your local agent should be able to let you have details but, failing that, contact the main agent or manufacturer. There is a world of difference, though, between that shiny new engine on a bench you worked on and the dirty, rusty beast living below the cockpit sole, so having learned the basics how do they apply to your engine? There may, be detail differences in design. The injector pump that was so easy to get at may, with yours, be obstructed by an engine bearer or you may find you can't reach a fuel filter without first emptying a cockpit locker. Familiarisation with your own boat is as important as inital knowledge and yet neither is any use unless you practise.

If the first time you have to bleed the fuel system or change a pump impeller is in the middle of the North Sea in a blinding rain squall with a bank less than a mile to leeward, your chances of success are low. Far better to practise, in moderation to prevent artificial wear and tear (not to mention creating fuel and water leaks), all the common repair/service routines on your mooring so when you have to do it for real you will know which spanner to use, which bolt to slacken first and discover for yourself any likely hitches.

But no practice or maintenance will help if engine access is poor or the engine difficult to inspect. Ian Nicolson's illustration shows ways of improving access and he also suggests that engine compartment lighting – one lamp on each side of the engine and a third over the stern gear – is more than a luxury as being able to see clearly what you are doing will make any task much easier.

Spares

The list below covers the most likely common failures and is realistic for normal cruising. For long-distance cruising many additional spares would be needed, if only to cover for problems with availability.

- Fuel filters (2)/spare glass bowl for water trap
- Oil filter
- Injector(s)/spark plugs
- Engine and gearbox oil
- Stern tube grease/oil
- Drive belts
- Water pump impeller
- Hoses and Jubilee clips

Additional items for petrol/outboard engines
- Coil
- Distributor cap and HT leads or, for breakerless ignition, appropriate ignition component
- Propeller shear pin

It should not be forgotten that tools and spares need maintenance as well as engines. Tools are often of mild steel and, unless well protected, will rust in short order to the extent that they can become unusable. The same goes for spares which, if neglected, can become more deserving cases for treatment than the component they are designed to replace.

Pre-passage check-list
- Fuel level and spare can
- Oil level, engine and gearbox
- Oil and water leaks
- Fresh water cooling – coolant level
- Drive belt for wear and tension
- Water trap
- Operate and check level in stern tube greaser
- Battery electrolyte level, terminal connections and voltage
- With engine running:
  cooling system – check exhaust for water flow
  listen for unusual noises
  check visible water piping for leaks
- Check alternator charge rate if ammeter fitted

Cracks like this at the base of the P-bracket are a clear sign of movement. It will be necessary to completely remove the bracket before re-fitting it with internal reinforcement

If the engine won't start or stops without reason
- Is there adequate fuel?
- Is there enough battery power?
- Check water trap and drain if necessary
- Increase throttle setting and try again
- Turn over engine for 10 seconds with decompressors open and either close or try again
- Is fuel getting through? Check flow by loosening inlet junction at the fuel pump
- Bleed the fuel system
- Check air inlet for obstructions
- Replace fuel filter(s), bleed the fuel system
- Is the engine overheating? Check drive belt and pump impeller
- Check exhaust for water flow

Familiarisation and practice have the added benefit that, at the same time, you may notice that piping is chafing through or that water is weeping from one of the junctions and do something about it. So if you've had the engine properly serviced, learned about the way it works and what is likely to make it stop and carried out your pre-passage checks, you're far less likely to have any problems. And the engine will probably appreciate all this extra love and attention and reward you by working when you need it most.

Chapter 8

The Fastnet – 10 years on

What is a gale? On the Beaufort Scale it is defined as being winds in excess of 33 knots. Many yachtsmen claim to have been at sea in a gale when winds temporarily gust over 33 knots but when in fact the *mean* windspeed at the time is no more than a Force 6 or, perhaps, Force 7.

In his introduction to the classic *Heavy Weather Sailing*, Adlard Coles describes a true gale as being *'when winds with a mean of Force 8 (33-40 knots) are sustained for an hour or more'*. Because the wind is never constant in strength, that 'mean' figure covers a range of windspeeds that will probably vary between Force 6 and 9 – even Force 10 temporarily in squalls – so the yachtsman unlucky enough to be at sea in such conditions may well record these squalls because it is they that cause the discomfort and damage. Adlard Coles goes on to point out that a true gale of this nature is rare in British waters in summer months.

I believe, however, that all these figures are arbitrary and that, for the yachtsman, a gale is a subjective judgement. A 25-footer in a Force 6 with wind against tide may be fighting for her life but in the same wind a heavy displacement 50-footer may be thinking of making her first major sail reduction and the crew of a supertanker beginning to feel the first lift of the sea.

To borrow a phrase from designer David Thomas, one of the members of our Offshore Yacht Advisory Panel, I would define a yachtsman's gale as being conditions when wind and waves are of such severity that seaworthiness and safety pass largely into the hands of the crew – in other words, conditions when the inherent strength, stability and sailing ability of the yacht are, alone, insufficient and the crew need to take positive action to ensure her safety. The point at which this happens will vary enormously from boat to boat, even between boats of the same size. The following articles on stability and ability of a yacht to make to windward in heavy weather identify those design attributes that go towards defining a particular yacht's 'gale' threshold.

Throughout this book the Fastnet storm of 1979 has been used as a yardstick. Conditions during that storm were of extreme severity and rarity and had it not been for the Fastnet Race it is unlikely that many yachts would have been at sea in that area. Even though, in August 1979, the forecasters were caught on the hop because the storm moved and intensified with such speed, major weather systems are usually predicted accurately and forecasts issued well in advance so that there is more than adequate time for yachts to reach shelter. Those choosing to stay at sea in the face of gale warnings take their own risks and must be prepared for the likelihood of survival conditions. Likewise, those venturing so far offshore as to be unable to reach shelter must be crewed and equipped so as to be able to ride out a full gale. But for the vast majority, even if a gale is announced as 'imminent' (within six hours) – and it is hardly likely to be a gale that has suddenly appeared out of the blue – shelter should be attainable and the gale ridden out in the security of a sheltered harbour or anchorage.

Many highly experienced yachtsmen have cruised successfully for years without ever having been at sea in a full gale. The long-distance sailor takes his own risks but, in the course of normal coastal and shorter offshore passages, it would be many lifetimes before the prudent yachtsman met anything near to those conditions encountered in sea area Fastnet in August 1979.

Andrew Bray

ON THE NIGHT of 13/14 August 1979 a storm of unseasonal and unusual intensity swept rapidly across the Celtic Sea. It was a storm that might have passed anonymously into meteorological records had it not been for one very significant factor. In its path lay 303 yachts taking part in the Fastnet Race. Fifteen people were lost in the storm, five boats were never recovered and scores suffered significant damage.

In the following pages we look back at that fatal storm and the casualties that resulted and we examine the seaworthiness and stability of today's cruising yacht in the light of the subsequent Fastnet Inquiry report and more recent studies. Additionally, we present the results of our own survey into the extent to which recommendations made in the **Fastnet Report** and **ORC Special Safety Regulations** have been adopted by production builders and of tests we have carried out on the strength of harness lifelines, hooks and attachment points

The Storm – a new theory

Former professional forecaster, ALAN WATTS, who carried out much of the original research into the Fastnet weather, has re-examined the evidence and proposes an entirely new metoerological phenomenon – the cyclonic pool

NO YACHT RACE in history has ever come close to rivalling the media attention that the 1979 Fastnet attracted. It was the horror of the loss of life that made that ill-starred race so reportable through the world's press. Books about it have proliferated in many languages, but there are still things we can only guess at.

When the weather over the sea builds towards its worst only a few isolated observers, who are not really in contact with one another, experience the conditions. They rarely get together and ask if there is any pattern to their experience. So the fact that one ship or boat experiences violent storm force winds, while others not far away are only visited by a simple gale of less, goes unremarked and unreported.

The Fastnet of 1979 was different. Here, in a survival storm, were experienced seamen who were all largely in the same club. With commendable attention to duty, a large proportion noted the conditions of wind, weather, barometric pressure and sea state in their logs throughout that night until disaster struck.

So, when the hastily contrived Fastnet Race Inquiry Report did not go anywhere near asking the right meteorological questions, K Adlard Coles decided to send out his own questionnaire. He was adding chapters to his classic *Heavy Weather Sailing* and needed precise information. As I had written the meteorological chapter for the original edition, Adlard asked me to frame the met questions, which I was very happy to do. As a result I was put in a unique position.

Never before – or since – has a sailing meteorologist been presented with reliable barometric pressures, windspeeds and directions, plus descriptions of sea state from a line of yachts in survival conditions. I hoped I could make some sense out of it all. That has not proved to be the case, but even

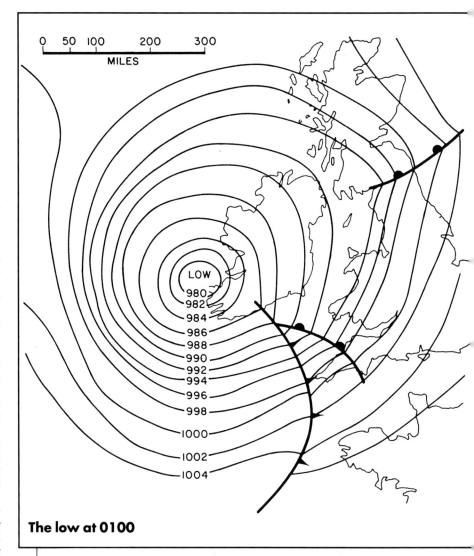

The low at 0100

Fig 1 The Fastnet Low at 0100, Tuesday, 14 August 1979. The isobars are at 2mb intervals and they are drawn without regard for any local pressure anomalies that were present

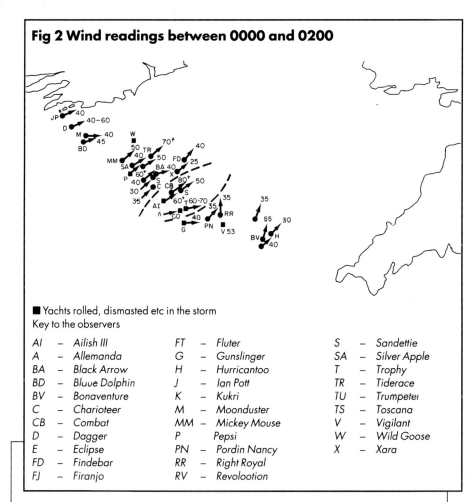

Fig 2 Wind readings between 0000 and 0200

■ Yachts rolled, dismasted etc in the storm
Key to the observers

AI	–	Ailish III	FT	–	Fluter	S	–	Sandettie
A	–	Allemanda	G	–	Gunslinger	SA	–	Silver Apple
BA	–	Black Arrow	H	–	Hurricantoo	T	–	Trophy
BD	–	Blue Dolphin	J	–	Ian Pott	TR	–	Tiderace
BV	–	Bonaventure	K	–	Kukri	TU	–	Trumpeter
C	–	Charioteer	M	–	Moonduster	TS	–	Toscana
CB	–	Combat	MM	–	Mickey Mouse	V	–	Vigilant
D	–	Dagger	P	–	Pepsi	W	–	Wild Goose
E	–	Eclipse	PN	–	Pordin Nancy	X	–	Xara
FD	–	Findebar	RR	–	Right Royal			
FJ	–	Firanjo	RV	–	Revolootion			

Fig 2 *The replies to K Adlard Coles's questionnaire were broken down into two-hour periods. These are the strongest winds recorded between midnight and 0200 Tuesday and they illustrate the degree of variation that was found. The zone within the dashed lines has winds that are in general 60 knots or more while relatively light winds are found just outside it*

the nonsensical bits are a step forward in our knowledge of what may actually happen at sea in a big summer storm.

When the replies came back, I found that I had reliable barometer readings from some twenty yachts. As they were given hour by hour, together with the times they were passing Scilly, I was able to use the known barometric pressures from St Mary's to check that the barometers and barographs were reading correctly, or to correct them where necessary. This way I could believe what they told me.

Only because of that was I able to plot the curves of pressure along the rhumb line Scilly to Fastnet Rock that we see in Figs 3 and 4 and truly believe what I saw.

As a forecaster, once professional but now amateur, I have always been brought up on the stock answer to how the wind develops as a depression comes in to the north-west of you. A broad wind field develops which more or less gradually increases in strength and will most probably come from around south somewhere. When the warm front has gone, the wind will now swing to a fairly steady southwesterly. Draw the isobars as the Met Office did for that night (Fig 1) and, because they have no pressure observations other than those ashore to tell them otherwise, they drew the isobars smoothly and the right distance apart to account for the windspeeds.

That was what I did also when I wrote an article (*The Storm*) for the October 1979 issue of *Yachting World*. Because they had no other source, the RORC and the RYA included that article with the official report, but I did it in advance of having real information about the pressure patterns over the fleet and the actual situation proved to be much more complex. Fig 1 is the situation at 0100 and it shows that, while the winds were still from the south-west, it was considered that the cold front had passed some time before. However, the very small kinks across the cold front indicate that, at surface level, there was not much difference between the air ahead and that behind the front.

The Low, over Shannon by that time, was certainly deep by summer standards and the isobars said that the surface winds ought to be a mean of about 40 knots (gale force) with the possibility of some gusts to maybe 60 knots.

Already, as early as 2200 the previous evening, there had been 60 knot mean winds with gusts to an estimated 75 knots in the area we shall have to refer to as the Labadie Bank. However, this bank is vastly too deep for it to have had any effect on the sea conditions in the area.

David Powell of the OOD 34 *Moonstone* told me, 'At about 8pm, when 60 miles from the Rock on the direct course, we noticed the clouds at 1,000ft were travelling diametrically against the surface wind direction of north of west 35 knots.' Now I submit that, for any normal meteorological situation, that is well-nigh impossible. You can get such contrary winds when, for example, the latter are light and sea breezes blow in under an offshore wind, but not when the surface wind is 35 knots and it is late evening out on the open ocean. There was something strange going on.

This is confirmed by Rodney Hill of the Oyster 39 *Morning Town*, which acted as radio relay boat. 'We were at one time,' he said, 'in the centre of something very strange, a swirling cloud all round and straight up clear night sky. There was mist and murk everywhere. Flares that we fired off went straight up through the 'hole'.' It would seem then that the flares were carried up a 'chimney' of rapidly ascending air and it is important to realise that, at the time, he was in the centre of the 10-mile radius circle within which nearly all the casualties occurred.

The German Admiral's Cup yacht *Jan Pott* reported, 'For 20 minutes from 0030 to 0050 the sky was clear, except approximately 20 degrees from the horizon. Fastnet Rock light was clearly visible distance five miles. Thereafter it disappeared completely and was never seen again, so at 0138 we had to tack just short of the Rock for safety reasons. The wind dropped by 15 knots during that time.' When about 30 miles on the way back *Jan Pott* was rolled through 360 degrees and dismasted. She came home on a mixture of emergency rig and engine.

It was this kind of observation, plus the checked but 'impossible' barometer readings from the yachts, that led me to an idea of how the rogue waves of that night were generated. I was aided by the results of work done by NOAA scientists as a result of four of their 'uncapsizable' Discus weather buoys being capsized by storms with characteristics very like those of the Fastnet storm. In all four cases it was found that the same strong winds occurred to the south of a deep low, but that a cold trough with intense convection cells and thunderstorms followed in behind. However, while the Rodney Hill observation looks like a kind of convection cell, no one reported any thunder during the Fastnet. There certainly was a trough and it came through the area long after most of the trouble had ensued.

Whatever the wind did, it was the seas that sparked the most awe and induced the worst effects. The Dutch skipper of Croix du Cygne said, 'Suddenly a 10 metre (30ft) high wall of water rose vertically over us. Then a roar and I found myself hanging by my lifeline in the water aft on the starboard side.'

That the, at times, 50ft waves were not everywhere is confirmed by the American 46ft *Aires*. They rounded the Rock about midnight and by 0200 were being swept by breaking waves every 10 minutes. The further east they sailed, the rougher it became. By 0500, when 50 miles from the Rock, *Aires* was being swept every minute. As they surfed down waves under storm jib, the mast and rudder were in violent vibration and a running backstay came away.

The crew of the 77ft Australian boat *Siska* had never seen anything worse even though they had sailed in winds of 50-60 knots before. The tops of the waves broke and toppled because wave research has shown that the crests in these windspeeds travel considerably faster than the bulk of the wave and so curl over like Pacific breakers. In an article in *Yachting*, Chris Bouzaid, the helmsman of the Australian *Police Car*, said, '... every sea was different. Some of them we would square away and run down in front of. Others were just too steep to do this. One imagines a sea to be a long sausage-like piece of water moving across the ocean. However, this was not the case at all as these seas had too many breakers in them and were not uniform.' The winds reported between midnight and 0200 are shown in Fig 2.

The non-uniformity of the seas is confirmed by John Ellis of the 32ft *Kate* (quoted in John Rousmanière's *Fastnet Force 10*), who wrote, 'The frequency was much too short for comfort, giving steep wall-sided seas, streaked heavily with foam and breaking (at worst) in the top 6ft. What seemed to happen was that three or four seas would build up into a succession of breakers, although these would not break along their whole length, but in isolated cusps which would then break outwards behind the fast-moving crest. At no stage did we see the Force 11-12 type of spume.'

We could go on quoting examples *ad infinitum*, but let us bring some facts together. By plotting the corrected pressures of about twenty boats (Fig 3 & 4), it was found that by 2100 GMT (Z) there was a low pressure area close to the Labadie Bank which was no less than 8 millibar below that to be expected (dotted line). This remarkable result is confirmed by three boats, *Black Arrow*, *Firanjo* and *Zara*, while the general pressure level is fixed by *Hurricantoo*, *Right Royal*, *Revolootion* and *Kukri*. These latter boats had relatively light winds compared to the 60 knots with gusts to 75 experienced by the boats in the depths of the 'trough'. This is remarkable enough, but this so-called trough had been there *in the same place since at least 1800Z and was associated with stronger than average winds at that time. It seems that, as the trough deepened, so the winds near it increased. By midnight GMT* the trough we are calling (A) had moved east and declined, but others (B) and (C) were detectable further west. By sheer bad luck *Black Arrow* again coincided with the depths of (B) while *Revolootion* and *Xara* lay between troughs and reported lesser winds at this time. By four in the morning (0300Z), (A) had disappeared, (B) was all but gone, while (C) was deepening and a further one (D) appeared near the Rock.

To me these observations are still a matter of immense wonder. I have re-checked them in the hope of finding I have made a mistake, but the records stand up. Undoubtedly a meteorological phenomenon unknown to science appeared over the Fastnet fleet that night. Rodney Hill must have gazed up into one of these strange chimneys and his flares were lifted up by rapidly ascending air currents. It was like a hurricane's eye, but even small hurricanes cannot exist in the main wind field of a depression whose centre is 100 miles or so north-west. Were there tornadoes over Fastnet? Tornadoes do not sit fixed to one spot for hour after hour, gradually develop and gradually decline. We would expect any such disturbance to have

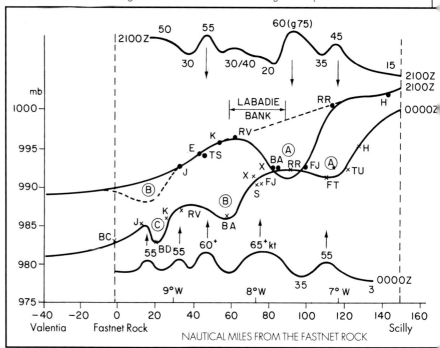

Fig 3 *To illustrate the strange pressure pattern, the checked and corrected barometer readings from nineteen yachts were plotted at three hour intervals. At 2100 GMT 'pool' (A) was no less than 8mb below the general pressure level (dotted line) while the existence of (B) is postulated being ahead of Jan Pott – the leader of this group. By midnight (A) had declined and moved away eastward but (B) and (C) had developed. The curves at top and bottom show the general variation in the reported winds at these times and illustrate that the strongest winds tended to occur on the edges of the pools*

Key to observers					
BA	– Black Arrow	FD	– Finndabar	RV	– Revolootion
BC	– Battlecry	FT	– Fluter	S	– Sandettie
BD	– Blaue Dolphin	H	– Hurricantoo	TR	– Tiderace
BV	– Bonaventure	J	– Jan Pott	TU	– Trumpeter
E	– Eclipse	K	– Kukri	TS	– Toscana
FJ	– Firanjo	PN	– Pordin Nancy	X	– Xara
		RR	– Right Royal		

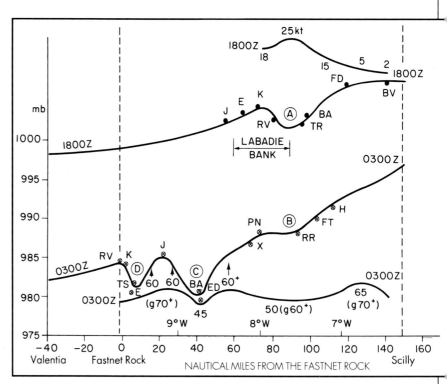

Fig 4 *To avoid overloading Fig 3 the pressure curves for 1800 and 0400 are plotted here. They show that pool (A) was developing as early as 1900 Monday evening and again the wind was strongest around it. However, the amazing thing is that it did not move for the next three hours or more. By 0500 the wind field had become much more uniform overall and in general the speeds were some 10 knots higher than six hours earlier*

(Figs 3 and 4 are from *Reading the Weather*, Watts (Adlard Coles Ltd))

been whisked away by the general run of the wind in which it was embedded. It seems impossible but the strange winds and the strange waves can be explained if we postulate a new met entity – a low pressure area that develops in the circulation of a major low BUT DOES NOT MOVE. For want of a better title, I have coined the name 'cyclonic pool' for this oddity.

To have such a low pressure, we have to have what is called *divergence*. That means more air flowing out of the air column over the low than enters it. If air was being sucked out of the area (A) faster than it was blowing into it, then the 8mb drop can be accounted for. Yet that inflowing air, sucked in to try and make up for that which was being lost aloft, had to be an addition to the wind that was already there. So where the two added 40-50 knots could become 60-70 knots and where they were opposed the wind could have been just 30-40 knots. This is just what was reported and when we plot the wind observations we see a pattern emerging. The strongest winds reported were on the edges of these cyclonic pools and as the pools grew, moved and declined so an intensely variable field of wind cast its net across the fleet, battering some craft, while leaving others virtually unscathed. We see now how it was that wind and waves were so extremely variable.

Waves are produced by the frictional force between wind and water, to which we have to add the sheer impetus of the wind on the back of high waves. It is usually assumed that the gain of potential energy through increasing wave height is at the expense of the kinetic energy of the air and so must depend on the square of the windspeed. This means that, theoretically, a 40-knot wind can raise a sea that is four times higher than a 20-knot wind can, while a 70-knot wind can raise a sea that is twelve times higher than the 20-knot wind can. These conclusions are borne out by measurement. However, the rate of working (the power) of a wind of 70 knots compared to one of 20 knots is over forty times greater. Thus the very strong gusts generate big waves very quickly. We do not need hours of duration as required by lesser wave fields for the waves to grow to maximum height. Here we are talking of building local giant waves in an already massive wave field when isolated corridors of excessively strong winds are generated within a lesser wind field.

We can imagine the cyclonic pools sucking in swathes of hurricane force gusty wind and, under the latter's impetus, monster waves being generated very quickly. But these waves are bigger than their surroundings and so break outwards across the boundary between the very strong and the not so strong winds, creating chaotic wave conditions – wave trains that are not 'sausage-like pieces of water moving across the ocean' as remarked by Bouzaid, but which would fit John Ellis's description admirably. George Tinley of *Windswept* can have the last word here. 'There were seas coming at one angle with breakers on them, but there were seas coming at another angle also with breakers, and then there were the most fearsome things where the two met in the middle.'

Andrew Beasley

Storm damage

Ten years after the Fastnet Race Inquiry Report was published, it still remains the most comprehensive analysis ever carried out into the effects of severe weather on yachts at sea and the lessons learnt from it are every bit as relevant as they were in 1979. The following presents extracts from the report relating to knockdowns and damage suffered by yachts in the Fastnet fleet

Knockdowns

The Report defined knockdowns as B1, which is to, or nearly to, the horizontal and B2, which is beyond the horizontal including a 360 degree roll. It then analysed those knockdowns in relation to size of boat and certain design characteristics. The boats are listed below in order of Fastnet Classes 0 to V according to their IOR ratings which equate approximately to the waterline lengths of the boats. The survey was based on a sample of 235 boats.

Incidence of knockdowns

Forty-eight per cent of all yachts suffered B1 knockdowns. Generally, the smaller the yacht, the greater the percentage* as below.

Class	0	(rating 42.1-70)	38%
Class	I	(rating 33-42)	28%
Class	II	(rating 29-32.9)	35%
Class	III	(rating 25.5-28.9)	54%
Class	IV	(rating 23-25.4)	54%
Class	V	(rating 21-22.9)	64%

Thirty-three per cent of all yachts suffered B2 knockdowns and again the smaller classes fared the worst.

Class	0	0%	
Class	1	15%	(8% of total B2s)
Class	II	10%	(5% of total B2s)
Class	III	46%	(31% of total B2s)
Class	IV	43%	(26% of total B2s)
Class	V	47%	(29% of total B2s)

Knockdowns in relation to design characteristics

The incidence B2 knockdowns related to the design of the boat gives an indication of what features might increase or decrease the likelihood. Those given here are the ballast, length/displacement, length/beam and beam/depth ratios.

Ballast ratios (ballast weight / displacement).

Less than 35%	50% suffered B2s
35% to 40%	31% suffered B2s
40% to 45%	43% suffered B2s
45% to 50%	33% suffered B2s
50% to 55%	30% suffered B2s

L/D ratios (displ / $(0.01L)^3$ x 2240). L is the rated (WL) length in feet and D the displacement in pounds.

Below 150	73% suffered B2s
150 to 174	31% suffered B2s
175 to 199	33% suffered B2s
200 to 224	30% suffered B2s
225 to 249	38% suffered B2s
Over 250	No B2s reported

Length/beam ratios (L / B), where L is the rated length and B the rated beam. A low figure indicates a wide beam but, as this tends to be more of a feature of smaller boats, in any case more vulnerable to knockdown, the significance of these

*These percentages may not be due only to size but also to the location at the time of the smaller boats which some reports consider experienced the worst weather.

figures should not be over-rated.

Below	2.5	61% suffered B2s
	2.6	39% suffered B2s
	2.7	35% suffered B2s
	2.8	31% suffered B2s
	2.9	36% suffered B2s
Over	2.9	No B2s reported

Beam/depth ratio (beam / centre mid-depth immersed). This gives an indication of wide, shallow hulls (high figures) which, like the length/beam ratio, is more likely to be a feature of smaller boats. The centre mid-depth of a boat is the depth of the 'canoe body' below the waterline – ie the hull itself without the keel.

Below 4.99	10% suffered B2s
5 to 5.99	24% suffered B2s
6 to 6.99	40% suffered B2s
Over 7	71% suffered B2s

The Report states that *'In the sea conditions experienced, characteristics which appeared to increase a yacht's likelihood of suffering a knockdown past 90 degrees include: lack of initial stability as indicated by high tenderness ratio (found by measuring the force needed to heel the boat to one degree) and low negative screening value (calculated from the tenderness ratio and other hull measurements to ensure the yacht is self-righting at 90 degree heel); wide beam as indicated by low L/B ratio (there is only slight indication that this factor was significant); wide shallow hull form as indicated by high B/CMDI ratio. There is little indication of any relationship between ballast ratio or length/displacement ratio and vulnerability to knockdowns.'*

Knockdowns and damage
Of the boats that suffered B2 knockdowns, the following is an analysis of damage caused*.
Base: 77 boats
63% suffered significant damage of which:
- 16% were dismasted
- 8% had minor rig damage
- 9% had floor damage
- 8% lost/damaged hatchboards
- 3% lost liferafts
- 6% damaged windows (NB there are restrictions on window sizes under ORC regulations)
- 6% suffered accommodation damage
- 5% had damaged steering

Damage to all yachts
Of the whole sample of 235 boats damage suffered breaks down as below:
a Significant rig damage 18% (mainly classes III,IV,V)
b Steering gear damage 11%
c Accommodation and interior 13% (mainly classes III,IV,V)
   Loose batteries and cookers were singled out as particularly dangerous
d Hull and structure (inc hatches and companionways) 14%

It is significant here that most of those (thirty-four boats) that reported under this category did so with reference to ancillary hull equipment rather than damage to the main structure of the yacht.

Actual hull/deck damage was reported

on just six yachts (18 per cent of damaged yachts, 2.5 per cent of full sample) and there was no correlation between displacement and structural damage. In view of the fact that these were mostly racing yachts and so, by definition, more lightly built than the average cruiser, the implication is that the structure of the production sailing cruiser would have withstood the storm without major structural damage.

Of those that reported significant damage, 21 per cent had damaged or lost washboards.

Abandonments
Twenty-four yachts were abandoned and only five were not recovered; of these one sank while under tow. With one exception, all abandoned yachts had suffered B2s. Seventeen abandonments were not made until help (helicopter, ship, yacht) was at hand. Only six were abandoned before help was on hand and, of these, two were not recovered. Two of those recovered had suffered knockdowns and sustained significant damage. Only two were abandoned on the premise that the liferaft was likely to offer more security than the virtually undamaged hull of the yacht.

In the conclusions of the report, the Committee made several statements which put the storm, the damage suffered by yachts and the casualties into perspective with regard to any perceived design or structural shortcomings of the boats taking part. We quote from these below.

'In every case there were a number of contributory factors...the common link between all fifteen deaths was the violence of the sea, an unremitting danger faced by all who sail.'

'It would be well to recall that the conditions experienced in the height of the storm, whilst no doubt precedented, must be regarded as an exceptional experience for most yachtsmen other than those engaged in very long distance sailing and in other waters than the south of the British Isles.'

'In the 1979 race the sea showed that it can be a deadly enemy and that those who go to sea for pleasure must do so in the full knowledge that they may encounter dangers of the highest order. However, provided that the lessons so harshly taught in this race are well learnt we feel that yachts should continue to race over the Fastnet course.'

Although these figures make disturbing reading, perhaps some comfort can be taken from one fact. Lives were lost, crew were injured and boats damaged and yet, after the storm, 298 of 303 yachts were still afloat.

Andrew Beasley

The reports of damage to yachts were made soon after the race. There is some evidence to suggest that damage was more widespread than initially reported with defects only becoming apparent when boats were later surveyed.

The stability factor

'Are you sure it won't tip over?' How many times have you heard those words from beginners and how many times have you replied 'Of course it won't – not with a couple of tons of cast iron stuck on the bottom'? The truth is somewhat different. Just as every boat has a range of positive stability – the range over which she will, when heeled, right herself – so she has one of negative stability, when she will prefer to remain inverted. Here we take a look at factors affecting stability

THROUGHOUT THIS series we have attempted in various ways to define the term 'seaworthiness'. It is not a concept that permits easy definition and cannot be explained in simple terms nor by a straightforward formula. Instead it is the combination of a large number of small parts that creates the whole seaworthy yacht. But it is a great deal more complex even than that, because each of those parts will in itself be the result of some form of compromise – a balance between function and practicality. If there ever was such a craft as the ultimate seaworthy yacht, the trade-off in other areas would be so great that she would be totally impractical to sail in normal conditions and to live on board.

In the earlier articles in this series, we have examined some of the elements that contribute towards seaworthiness. We have studied the watertight integrity of hull and deck, rig and sails, deck layout and gear, interior accommodation and engine reliability. But when thinking about seaworthiness, the two factors that are probably uppermost in every yachtsman's mind are stability and the ability to make to windward in heavy weather. In this article we examine factors affecting stability.

What is stability?

Stability depends on the inter-relationship of two factors, centres of gravity and buoyancy. *You might* measure it in terms of the amount the gunwale of a boat dips when you first step on board or how far she heels when you are sailing. If you are a pessimist you might worry about it and wonder at what angle of heel your boat will prefer to turn upside-down rather than right herself. Every boat has a *stability range* which goes from that initial dip of the gunwale until she prefers to remain inverted. The righting moment of a boat – the result of that interaction between centre of gravity and buoyancy – will, if plotted against angle of heel, form a curve (the so-called GZ curve, see illustration) which passes through two critical points. One is the point of *maximum stability* and the other is the point of *vanishing stability*. But let's go back to the question of ballast (or c of g) and buoyancy.

An unballasted catamaran relies almost entirely on its buoyancy for stability. As pressure builds up on the rig so the windward hull lifts a little, the leeward one is depressed by the same amount and the buoyancy of the one is transferred to the other to counter the heeling moment. A Metre boat, a 12 Metre for example, relies almost completely on keel weight for

A capsize captured by photographer Lucille O'Neill. It is not in a gale; nor is it in UK waters. In fact it is not even the sea. The pictures were taken at the entrance to Michigan City on Lake Michigan in the USA. A gale had blown the night before but conditions had calmed and a number of yachts attempted to leave. Rollers were breaking in the entrance on account of shoal water so this type of breaking sea is not of the type encountered in deep water.

*The boat, **Bay Bear,** is a 42ft 1964 Sparkman & Stephens design, with a displacement of 18,960 lb, ballast ratio of 47 per cent and an angle of disappearing stability of 128 degrees. At the time she was on charter to the US Navy. As she approached the entrance a wave swept her deck and she lost steerage way. The following wave hit her and knocked her right over. She quickly righted and as all crew were harnessed on, even though several were in the water, all were saved. The damage was a broken finger, some bruises and some lost and broken loose gear (including the masthead wind indicator)*

stability. She leans over and it is the physical weight of lead in her keel that counters the heeling force.

A simpler analogy is the shoe-box and the pendulum. The shoe-box will have enormous initial stability but once it starts to tip it loses it very rapidly and gets to the stage where it capsizes and will lie capsized every bit as stable as it was the right way up. By contrast, the pendulum will always swing back from whatever position except absolute top dead centre. And then it needs a nudge of just a fraction of a degree and back it swings.

It is a popular misconception that the maximum stability of a yacht occurs when she is heeled at 90 degrees, when her keel is creating maximum righting moment. This is the case with the pendulum but then look at the shoe-box – maximum stability is at 0 degrees of heel. The pendulum would not, in any event, make a very satisfactory boat as it has no buoyancy. Introduce buoyancy and you begin to have a measure of *form stability*, that is stability generated by the buoyancy. And as soon as you have form stability you have a boat that will, however narrow the range, have a degree of inverted stability.

So these, then, are the goalposts for the yacht designer when considering stability. Form stability (the shoe-box boat) gives high initial stability that vanishes at a low angle whilst a low centre of gravity creates little initial stability but has a greater ultimate value. Somewhere between these two are the boats that you and I sail.

There are many other factors that affect this simplistic view of stability but, taken superficially, it would seem to suggest that we should all be sailing pendulum-type boats such as the Metre classes. These are very narrow, easily-driven boats with high ballast ratios; but their disadvantages are numerous. Although their vanishing stability is high, initial stiffness is low so in the normal range of sailing conditions they sail at high angles of heel. Because they are narrow accommodation is poor and because they have low buoyancy and high displacement they lift less to the sea, tending to sail through rather than over the waves which makes them very wet boats. And it is not unheard of for a boat of this type to be physically sailed under, an event far less likely to occur with the shoe-box boat.

If boats were box-shaped then stability calculations would be very simple. Unfortunately they are not. Their complex shapes give birth to a whole host of features that can each add or detract from the stability equation. These range from length and beam, hull depth (canoe body) below the waterline, displacement, ballast ratio, height and flare of topsides and even the coachroof can have a significant effect if the boat is knocked down.

The accompanying table shows the major factors affecting stability. Beam, hull depth and centre of gravity are the three key elements. Like the shoe-box, a wide hull will give high initial stability and one that will heel less over the normal range of sailing conditions. However, even if it has a low centre of gravity, that picture changes dramatically once past the point of maximum stability and on the downward slope of the GZ curve. When it reaches vanishing stability the beam, the asset that gave it high initial stability, will, for the same reason, become a liability and it will have a greater range of inverted stability than a narrow boat. And because wide boats tend to have shallow hulls and higher centres of gravity than narrow, the point of maximum stability will be lower. So it is the light displacement, wide and shallow-hulled yacht that is more prone to knockdown. And it is these features that have crept into some areas of cruising yacht design in recent years in the quest for more space (=beam), competitive prices (= light displacement) and performance (= low wetted surface and shallow hulls).

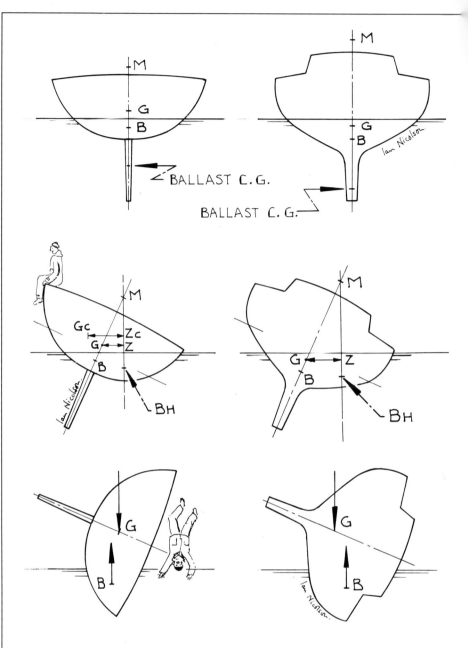

Understanding stability and the interaction of centres of gravity and buoyancy. In all three illustrations the left hand yacht is a shallow-hulled, lightweight beamy hull, the right hand one a more traditional shape.
Top: *when the boat is upright the centres of gravity G and buoyancy B are in the same vertical plane so the boat is in equilibrium. M is the metacentre, the centre of the circle defined by B as the boat heels. The wide, shallow hull has centres of gravity and buoyancy well above those of the traditional hull*
Centre: *As both hulls heel so the centre of buoyancy B moves to leeward and becomes BH (buoyancy heeled). G also moves but to a lesser extent and so a righting lever GZ is created. As can be seen, the shallow hull has a smaller GZ than the traditional until crew weight is placed on the weather rail producing a modified righting lever, GcZc, where Gc is the new centre of gravity allowing for crew weight*
Bottom: *Knockdown, with both hulls heeled to over 110 degrees. The shallow hull is beyond the point of vanishing stability as G has moved beyond Z and is now on its way to a complete 180 degree capsize. Crew weight no longer plays any part. The traditional hull, meanwhile, maintains a fair amount of positive stability. Note also the role of the high coachroof in improving stability at this angle*

How wide is wide...?

...and how light is light, how shallow shallow? The ratios which together start to build a picture of yacht stability are: ballast ratio

High initial stability (desirable for sail carrying)

For	Against
Low C of G	High C of G
Wide beam	Narrow beam
Heavy displacement	Light displacement

High maximum stability (for sail carrying)

For	Against
Low C of G	High C of G
Wide beam	Narrow beam
Shallow hull	Deep hull
High topsides	Low topsides
Large coachroof	No/low coachroof

Large angle of maximum stability (desirable for safety)

For	Against
Narrow hull	Wide hull
Deep hull	Shallow hull
High topsides	Low topsides
Large coachroof	No/low coachroof

Low inverted stability (desirable for safety)

For	Against
Narrow beam	Wide beam
Low C of G	High C of G
Large coachroof	No/low coachroof

Stability in perspective

To the owner of a standard production boat of moderate proportions, stability should not be a matter of any great concern. Experience gained during the 1979 Fastnet Race confirmed that small boats are vulnerable to capsize in large, breaking waves. I don't believe there is anything a designer can do that will alter that fact. He can reduce the chances of a boat being knocked over by increasing her displacement; he can increase the probability that she will right herself quickly by giving her a narrow beam, a large coachroof and plenty of ballast slung as low as possible. But, for a given size of boat, he can do nothing to alter the relative sizes of boat and wave, which is the essential cause of capsizes. If a small enough boat meets an exceptionally large, steep, breaking wave, the boat is likely to be knocked over.

Naval architects define stability in terms of the GZ curve. No doubt this approach gives a good indication of the adequacy of stability; it defines the range of positive stability, the angle of maximum stability and, when combined with the displacement it gives an indication of the energy needed to heel the boat to a particular angle. But because the GZ curve deals with the static situation, it cannot tell us how a yacht will perform in the dynamic conditions at sea.

There has been some work done on dynamic stability, both in this country and in the US. Whether this work is really useful is doubtful. Very few yachts capsize, because most of us are sufficiently devout cowards as to avoid the conditions in which a capsize is likely. Thus we do not regard the yacht's roll moment of inertia or lateral area of keel (in a scientific as opposed to subjective sense) as key features when we decide what yacht we are going to buy. There is no strong market force which will give the designers or builders the incentive to do a great deal of research into a new design to ensure that she is, as far as possible, proof against 120 degree knockdowns or 360 degree rolls.

There are a number of empirical screens which will identify potentially dangerous boats, the RORC SSS screen being probably the best known. This approach establishes 'normal' dimensional ratios, gives credit for anything likely to be safer than the norm and penalises anything more dangerous, which seems sensible. The difficulty with it is to identify the dividing line between acceptable and unacceptable but, as a system of monitoring and discouraging dangerous trends, it has merit.

Stability is not just about a boat's vulnerability to capsize. It is also about her power to carry sail. This is the aspect of stability that has suffered most under the IOR rating rule. The rule makers were keen to encourage boats with a reasonable amount of accommodation, to make sure that the weight of furniture and fittings was not a penalty. Thus they wrote the rule in such a way that the boat with a stripped out interior, and hence a large percentage of her weight in the keel, would be penalised.

Then the designers got hold of the rule and produced boats with light, stripped out hulls and even lighter keels. They also produced beamier boats and the power to carry sail depended on the crew sitting on the weather rail – not a useful trend for cruising boats.

The racing hull shapes of the last two decades have been highly specialised. The lines are distorted to make the boat seem, from her measured dimensions, to be slow. She is designed to be sailed upright and this can only be done by putting a great deal of crew weight in the right place, carrying the right sail and steering accurately. As long as she is treated as intended, she is an excellent performer. Treat her wrong and she is a pig to sail, lacking stability in every sense.

Bill Anderson

(ballast weight / displacement as a percentage); length/displacement ratio (displacement in lb / [0.01 x LWL]^3 x 2240); length/beam ratio (LWL / beam); beam/depth ratio (beam / maximum hull depth [ie hull without the keel]). So if a boat, compared with others, scores low on ballast, length/displacement, length/beam ratios and high on beam/depth, she is likely to have a lower angle of vanishing stability.

But it *is* still only an indication because such simple formulae cannot attempt to define the complex curves of a yacht's hull nor her centre of gravity. She might have a low ballast ratio but if that weight is carried low in the keel in a bulb then her centre of gravity may be substantially lower than a yacht with a higher ratio.

The only sure way to establish the static stability of a yacht is to measure the forces needed to heel her to one or two degrees in an inclination test and then with accurate hull data, either from lines drawings or by measurement, have an establishment such as Southampton University's Wolfson Unit calculate a stability curve.

But, whilst such calculations are easily carried out with the right data available, the cost for an individual owner is not cheap and it raises the question of why such information is not more readily available from builders and designers. There is a strong argument that GZ curves and stability figures should be published as a part of the sales literature for every production boat to allow prospective customers to make sensible comparisons. If nothing else, it would give yachtsmen a better understanding of stability.

Trying to calculate and define stability in simple terms is a problem that has taxed many brains the world over. Racing authorities such as the RORC have a system of screening values which are used to set limits for various categories of race under the IOR and Channel Handicap. But these values are not based on stability alone; they take into account a host of other features which may affect a yacht's offshore ability. The accompanying table gives a few examples.

Recently, in the wake of the *Marques* disaster, the Department of Transport commissioned the Wolfson Unit to carry out a research programme into stability in order to

Ballast, beam, length and displacement ratios

Boat	Ballast Ratio	Length/Displacement	Length/Beam
Feeling 1090	41%	199	2.4
Moody 346	40.2%	229	2.4
Oceanis 350	37.5%	176	2.7
Rustler 36	45%	403	2.4
Sadler 34	39%	263	2.6
Victoria 34	43.4%	250	2.7
Voyage 11.20	34%	253	2.5
Westerly Falcon (Sea Hawk)	41%	319	2.2

All the above boats have been chosen on the basis of being of similar length overall. They vary from the long keeled Rustler 36 through to lighter and beamier boats. Predictably, the Rustler comes out with the highest figures for ballast and length/displacement ratios. Whilst none of the boats exhibits low figures in all ratios, if the same calculations gave the ballast ratio of the Voyage 11.20, the length/displacement ratio of the Oceanis 350 and the length/beam ratio of the Falcon for a boat it would be an *indication* of a low stability range but no more than that.

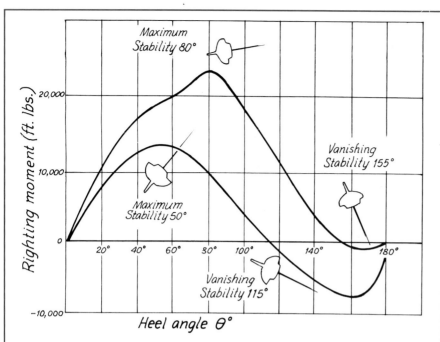

These curves show the righting moment of two hulls similar to those in the diagram on page 128. The righting moment is in ft lb and the reserve stability is represented by the area below the curves. Although these two examples are quite extreme they show clearly the differences in reserve stability and the vastly different points of maximum and vanishing stability

produce a series of screening values, based on simple measurement and calculation, for training vessels and sailing school yachts. The final recommendations, which will cover everything from equipment carried to fire prevention as well as stability, are expected to come into force next year. When these are published, the values and sailing limits defined should, if properly interpreted, start to give a clearer picture of the seagoing abilities of the modern yacht.

Stability isn't everything

Even the most meticulously calculated GZ curve only defines static stability. The trouble is that, in the conditions when a knockdown is most likely to occur, in heavy breaking seas, a boat is far from static and she is subjected to a multitude of dynamic forces.

For the same reason that the wide yacht rights itself when heeled 30 degrees in calm water, it will be heeled if the water surface is inclined at 30 degrees, as on the face of a wave. So form stability actually induces the roll that is the first stage of a capsize. Pendulum stability will still be operating in your favour, but there is a school of thought that suggests that a capsize resulting from the impact of a breaking crest will, when it does

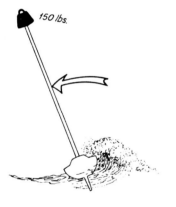

A heavy boat will have greater capsize resistance than a light one though her structure will need to be stronger to withstand the force of breaking seas as she 'gives' less. Likewise, whilst a heavy mast might normally be a disadvantage in that it decreases static stability (raises c of g) and will make the boat roll more, it will also increase capsize resistance because a greater force from a wave will be needed to roll it to a capsize (a high roll moment of inertia).

Channel Handicap screening values

Typical base value of SSS for a number of classes. These do not include the adjustment value and self-righting factor, which have to be applied for, so individual boats may have higher final SSS numerals.

Design	LOA (metres)	SSS base value
Sonata	6.55	15
Sadler 25	7.32	20
MG C27	8.38	19
Laser 28	8.66	16
HB 31	9.25	22
Contessa 32	9.66	35
X99	10.00	23
Storm 33	10.15	34
Hustler SJ 35	10.54	29
Swan 371	11.23	39
Sigma 38	11.55	40
Lightwave 395	12.04	30
Sigma 41	12.42	45
Nicholson 55	16.61	60

ORC categories of races with the minimum recommended SSS numeral for a yacht to take part.

Minimum SSS numeral	Category number	Outline of ORC description
10	4	Short inshore races in warm or sheltered water
20	3	Races in open water, mostly close to shore
30	2	Extended duration with high degree of self-sufficiency required, but outside assistances could be called (RORC races)
40	1	Long distance, well offshore with yachts self-sufficient for extended periods, without expecting assistance
50	0	Trans-ocean races, expectation of heavy storms, able to deal with serious emergencies without assistance

How heavy is heavy, how light is light?

In his book, *Seaworthiness, the Forgotten Factor* (Adlard Coles), Tony Marchaj defines different displacements in terms of the displacement/length ratio (displacement in lbs / (0.01 × LWL in feet)^3 as below

150 and above	Very heavy
400	Heavy
350	Medium heavy
300	Medium
250	Medium light
200	Light
150 and below	Very light

happen, be more severe for a yacht with a low C of G. This is because the yacht will tend to rotate about the C of G and the impact will, therefore, have a greater lever if this is low. So static stability, although highly desirable, is only one of the elements that will increase resistance to knockdown.

Size is the biggest single factor. A small yacht, however stable, will be more easily rolled than a large one. As a rough guide, if you double all the dimensions of a boat you increase her stability sixteen-fold and, as she is heavier, she has greater capsize resistance. In practice, this would produce a boat with such a violent motion as to be almost unsailable and most yachts, whatever their size, are designed with similar stability characteristics over the normal range of sailing angles because they are sailing in the same range of winds. It is only their displacement that gives larger yachts a greater righting moment. This is the fundamental reason why wide yachts have high centres of gravity. They do not need a low C of G in order to carry their sail at *normal* sailing angles.

Even on boats of the same size, displacement is important. Whilst it can be argued that a light boat gives more to the seas and 'rides the punches', it can equally be argued that with her lower mass she is less able to withstand the force of a breaking sea and so more likely to be capsized. In contrast, although a heavy boat might be more able to withstand a breaking sea and avoid capsize, she needs to have a far stronger structure to escape damage.

Weight distribution also plays a part. A stripped-out racing boat with all her weight concentrated amidships will have less capsize resistance than one where the weight is spread more evenly laterally across the boat. A boat with a heavy rig may roll or pitch more than one with a light mast but it will have a greater resistance to capsize on account of the rig's inertia – in other words the wave will need to have that much more energy to set the rig into motion towards the capsize position.

High freeboard is an asset through the buoyancy it creates when a boat is knocked down or inverted, whilst low freeboard can suffer a further disadvantage during a capsize in that she may 'trip' over her side deck as she is knocked down the face of a wave. And it is ironic that one of the major factors in preventing capsize, a low centre of gravity, can actually increase the chance of a 360-degree roll because of the pendulum effect. A deep, heavy keel can act as a flywheel during a capsize. As the boat is knocked down, even though she may have a high angle of vanishing stability, the sheer inertia of the keel as it rotates can be enough to complete the roll.

So far, we have not considered the most important element of the lot, the crew. In the opening piece, we described the concept of a yacht's 'gale threshold', a wind strength and sea state above which seaworthiness and safety pass largely into the hands of the crew.

It is in these conditions that the skill and common sense of that crew and the tactics they employ will make the biggest contribution to capsize avoidance. Whilst there is no such thing as the uncapsizeable yacht, stability, capsize resistance and the skills of her crew all add up to make a considerable contribution to her overall seaworthiness. ❐

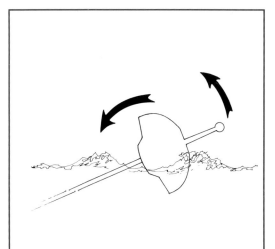

The **'flywheel'** *capsize. If a boat with a deep, heavy keel is knocked down, the inertia of her keel rotating may push her past the point of vanishing stability when a yacht with a lighter keel might recover*

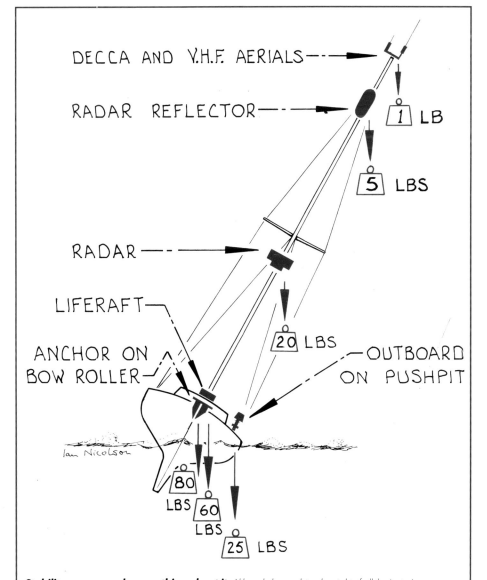

Stability – you can do something about it. *Although the combined weight of all the 'extra' equipment shown here is not that great at 191 lb, the lever effect it has is far greater. If this mast is 38ft long and the draught of the boat is 5ft 6in, she would need to carry a bulb weighing 235 lb at the bottom of her keel to achieve the same stability as a sister ship with no extras. Whilst all one can aim for with fixed items such as radar is to select a light scanner and bracket, why not have the reflector on a halyard and stow anchor, liferaft and outboard as low as possible?*

Windward ability – drive vs drag

Ever since sailing craft first took to the water, seamen have had an almost instinctive dread of the lee shore. It was the inability of vessels to make to windward in strong winds that led to this fear and yet, with efficient modern rigs and hulls, windward ability still plays a major part in the overall seaworthiness of a boat

GENTLEMEN, they say, never sail to windward. And for the rest of us who do, it must be the least favourite point of sailing. Heel angle increases, the motion of the boat becomes more lively as she sails into the seas, the wind into which we are sailing increases in apparent strength and contrives to throw spray back into the cockpit. But without windward ability a boat faced with a lee shore is little better off than a piece of driftwood.

In moderate conditions, the modern cruising yacht is a marvellously efficient sailing machine when going to windward. Even the pot-bellied bilge-keeler with baggy sails succeeds in making respectable ground to weather. It's only when the time comes to make the first sail reduction that the picture begins to change.

It is easy to believe that, in light airs, sail shape and efficiency is far more important than in heavy going when, with so much more drive available from the extra wind, the loss of rig efficiency is more than made up for by the surplus breeze. The opposite is the case. You may have more drive at your disposal but you also have more drag in terms of the windage of the non-drive parts of the hull and rig. The more the wind increases, the more sail is reduced (= less drive) but the greater the drag until a point is reached when that particular yacht can no longer sail to windward because drag is greater than drive and she makes so much leeway that progress to weather is halted or even reversed.

But that is not the whole picture because, as well as windage and rig efficiency, the same factors that affect stability have a very significant effect on a yacht's ability to make to windward and it is fascinating to see just how closely related these are.

Whilst stiffness (= high maximum stability) of a hull is desirable for windward sailing, the way that is achieved can have a major effect on windward ability. Closely related to stiffness is the way the boat handles when heeled and, although it does not affect a boat's inherent ability to windward, a boat that is difficult to steer upwind will ultimately be less efficient as she will tire her helmsman quickly and sail a less effective course compared with one that tracks well.

This is one of the biggest drawbacks of the modern wide-beamed and relatively

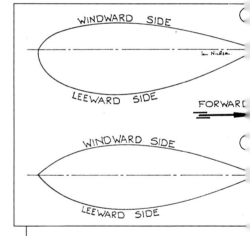

The effect of heeling on wide and narrow boats. As the wide boat (top) heels so her waterline shape becomes distorted, creating a tendency to broach to windward, an effect exacerbated by her wide stern causing the rudder to lift and stall. By contrast, the narrow hull (below) will tolerate greater angles of heel

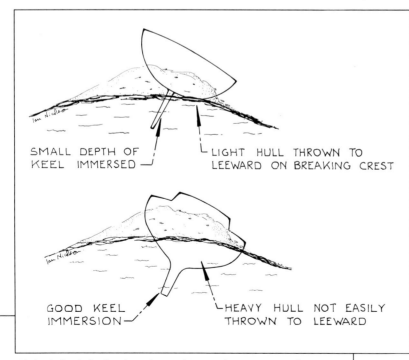

A light, shallow hull will tend to make more leeway in heavy seas as it is thrown to leeward on breaking crests. The heavy boat remains more fully immersed with a better grip on the water

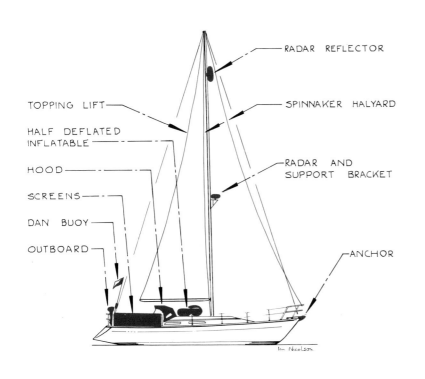

Improving windward performance. *The areas of the extra items shown here add up to about 35sq ft – all of which provide drag but no drive. The windage from these increases as the wind rises but at the same time drive from the sails lessens as they are reefed. This can be reduced considerably by stowing loose gear below decks, taking halyards and topping lift back to the mast and even, if progress to weather stops, removing the side screens and folding down the sprayhood*

shallow-hulled cruising boats. On account of hull form, they are comparatively stiff and will carry sail and handle well until heeled to about 20-25 degrees. At this point, because of the very distorted heeled hull shape (see illustration), and because, when heeled, they tend to ride on their 'shoulders' and lift their shallow sterns allowing the rudder to become aerated and begin to stall, weather helm builds very quickly until they break away and gripe to weather.

In squally weather, this makes for a boat that is 'twitchy' and difficult to steer well and the solution is to reef early so as to sail more upright. This means that, except in the gusts, the boat will be undercanvased and her ability to sail to windward reduced. The fact that this type of boat needs to be sailed fairly upright means that the efficiency of her rig will need to be greater as the sail area is more squarely presented to the wind; this again demands more skill on the part of her helmsman.

At the other end of the scale, the narrow and tender yacht also suffers. She might continue to handle well but a high angle of heel, 35 degrees or greater, makes her uncomfortable so she reefs to sail at a more reasonable angle. In high windspeeds this means that she too will sail undercanvased and start to lose drive well before a stiffer boat with greater sail carrying ability.

Improving windward ability

A more appropriate heading for this section might be 'preventing the deterioration of windward ability'. A boat straight from the factory with well cut new sails will perform very much better on a number of counts than one of the same class a couple of years older. The first is the condition of the sails themselves and the second is what the owner will have added to the boat to create extra windage and weight.

The accompanying illustration shows typical extras that might be added, from sprayhood to radar reflector and these add up to an incredible 35sq ft, more than an 8ft x 4ft sheet of plywood. That area is all dead, non-working windage and an appreciable percentage of the whole boat's windage. If you imagine holding that 8ft x 4ft sheet of ply at an angle of 45 degrees to a Force 6 wind then you'll have an idea of how much drag is being created and how much extra drive from the sails will be needed to overcome it all. It will also be adding to the heeling forces on the boat so you'll be working at a double disadvantage.

At the very least, extra halyards and topping lift can be taken back to the mast and dinghy, outboard and anchor stowed below deck. But when the going gets really hard and your progress to windward begins to slow to a crawl, then if the safety of the yacht is in question it's time to stow the side-screens, lower the pramhood and be ready to get wet. It could just make the difference between being able to sail away from that lee shore and not being able to.

Weight and weight distribution are also important. The more that heavy items are moved away from the ends of the boat, whether anchor, outboard or dinghy, the less the boat will pitch and the more efficiently the rig will be able to operate. So again, if a heavy weather windward passage is anticipated, a few minutes spent stowing heavy gear as near the centre as possible will pay dividends.

In the borderline situation, the lateral distribution of weight may also make the difference between making ground to windward and not. Heavy gear can be stowed in a windward berth, crew persuaded to sit or lie on the windward side of the boat, whether on deck or down below – every little bit will help.

Improving windward ability by taking positive steps can do much to improve a boat's inherent performance. In the end, if you've done all you can to help your boat, the deciding factor may be not the sails but the engine, as detailed in the last chapter.

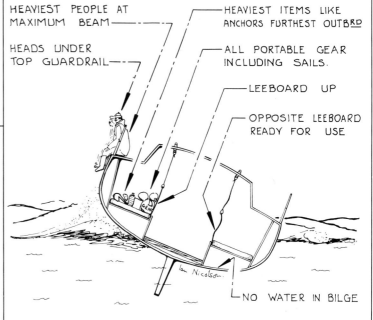

Although, in extreme conditions, it would be dangerous to have crew on the weather rail, for a short windward passage in strong winds the effect of their weight on stability will have a very marked effect on speed to weather. If faced with a mutiny, the least you could do is to ensure that they sit or sleep to weather and also transfer all heavy gear to the weather side of the boat – it's illegal when racing but quite practical for the cruising boat. If, however, the weather is extreme enough for a knockdown to be a risk, all heavy gear must be securely stowed to prevent the risk of injury from flying objects

Clipped on – but are you safe?

The diligent yachtsman always harnesses himself to the yacht when going on deck or in bad weather. But is this really strong enough; are there weak links in the average harness tether or its attachment? *Yachting Monthly* **conducted independent trials**

FEW YACHTSMEN who weren't at sea on 13/14 August 1979 can appreciate quite the localised spite of conditions of the Fastnet storm, but every one of us realises the vital part that a yacht harness plays in bad conditions. One of the sobering statistics to come out of the 1979 Fastnet race was that 11 per cent of yachts suffered harness failure of one sort or other.

That night yachts were being rolled every few minutes with crew being scattered in all directions as venomous seas picked yachts up and turned them over time and again. Harnesses saved hundreds of yachtsmen that night but still six of the fifteen deaths in the 1979 Fastnet Race were directly attributable to the failure of harnesses or their attachment points. Twenty-eight per cent of the fleet reported that, with hindsight, they would make changes to the points used for harness attachment.

These figures led us to conduct trials on a variety of methods of harness attachment to yachts, plus testing of two tether hooks.

We didn't examine harnesses themselves because the British Standards Institute produce a comprehensive standard, BS4224, to which they should be constructed and tested.

In reality, whilst most harness manufacturers claim to build in accordance with these standards, the expense of this formal testing makes the harnesses uneconomic, and generally the companies satisfy their quality control requirements with private tests. However, there are two commercial yacht harnesses, the Sowester and Ancra which are BSI approved.

The Fastnet Report recommends harnesses which comply with BS4224, and it goes without saying that *in extremis* any harness with a question mark hanging over its strength has to be regarded as a false economy, as was borne out in the Fastnet when only a tiny percentage of harnesses were Kitemarked.

The YM tests

From a logistical viewpoint, it was impossible for us to test every location to which a yachtsman might attach his harness hook. The variations are too numerous, from stanchion bases to guardwires or from standing rigging to jackstays. Instead we took typical, if substantial, stainless steel U-bolts and fitted them in a variety of different ways to a number of specially-constructed GRP panels. It is fairly obvious that a U-bolt attached with normal washers will not have the strength of one fitted with, for example, a substantial backing plate – but just *how much* weaker would it be?

The tests were carried out at Marlow Ropes' R & D Department using an hydraulic puller capable of delivering a smooth load of up to 30 imperial tons. Although the nature of harnesses in use gives snatch loadings, Marling Industries (who produce the most popular yachtsman's harness, the BSI-approved Ancra) find that there is no significant variation in either the mode or level of failure between slow and snatch pulls.

We employed a popular brand of standard 8mm (3/8in) thick 51mm (2in) 316 quality stainless steel U-bolt costing approximately £7.50 each. A stainless plate bridge is fitted, against which the nuts are tightened.

Eight 1ft (300mm) square GRP panels were constructed to represent as typical a lay-up as one might expect to find in the cockpit area of an average 35ft cruising yacht, four of them solid laminated GRP (comprising eight laminations of 1½oz = 12oz lay-up, approximately ¼in thick) and four of them balsa core sandwich (4 × 1½oz/¼in balsa/1½ oz/¼in balsa/3 × 1½oz, resultant being approximately 13/16in thick).

We tested four methods of attachment to both sets of panels. Fitting holes were drilled accurately and carefully in the glassfibre panels so there was no slackness. The panels themselves were fitted snugly into a steel frame for the purposes of the test.

Panel 1

The U-bolt was attached to this solid-glass panel with the small washers provided with the fitting. As pressure was put on, the panel bowed considerably until, at a little over 1 ton pull, the GRP could be heard gradually breaking down. At 1.66 tons the U-bolt suddenly pulled away drawing both washers and nuts cleanly through the laminate. The U-bolt was undamaged.

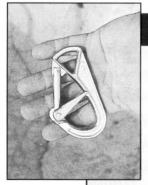
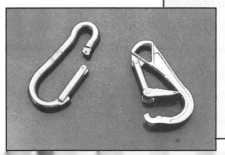

Hook tests

Although no actual harnesses were tested, we decided not to miss the opportunity of testing two typical harness clips – a standard carbineer-type, and the Gibb locking carbineer introduced when it was realised that, when capsized over an average U-bolt, a standard carbineer hook can very easily trip itself.

The two were similar sized, the standard 316 stainless steel carbineer costing £5.50 and the Gibb over twice as much at £12.70.

Each had specifications stamped on it – the carbineer 'Max 2000 – Use 450kg' and the Gibb 'Tested 11KN'.

The carbineer was fitted with a mechanical pin latch for the gate. It pulled steadily up to 2.2 tons before the hinge disintegrated allowing the arch to straighten slightly, releasing the load. We had expected the gate opening to give first, but the weakpoint appears to be the hinge.

The Gibb went to 2.9 tons before the arch gave way and straightened enough to release the load.

Both performed very well, and exceeded the specifications required from BS4224. In practice the Gibb is clearly the better of the two in being un-trippable, but then if proper precautions are taken (the carbineer cannot trip itself on wire, for example, or small-circumference eyebolts), a standard carbineer appears to be perfectly acceptable.

We would point out that there are a number of other similar hooks available, Crewsaver in particular making one very similar to the Gibb.

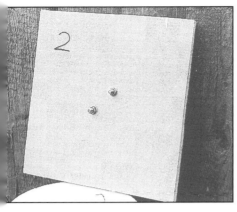

Panel 2
The U-bolt was attached in an identical manner to Panel 1, but this time to a balsa sandwich panel. As the load came on, the glassfibre started breaking down at 1 ton, with considerably less deflection than its twin Panel 1. The U-bolt pulled through gradually, implying a degree of shock absorption, with one arm of the U finally pulling through at 1.8 tons. The U-bolt was undamaged

Panel 3
This solid GRP panel had the U-bolt attached with 1 3/8in (35mm) 'penny' washers. The normal washers were fitted between the penny washers and nuts. Interestingly, after bowing the panel considerably, the loading that the two large washers imposed on the panel split it across at 1.8 tons. It still withheld 1.1 ton before final failure, the U-bolt releasing through the split.

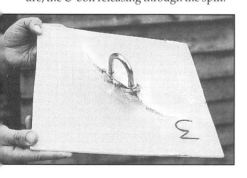

> ### Padeye tests
> Out of interest we also tested this stainless steel padeye which might be employed to attach harnesses to. The base plate measured 2¼in (57mm) square on to which was welded a 3/8in (8mm) stainless U-ring.
>
> For the purposes of the test we attached two of these back to back using 5mm pan head nuts and bolts (the holes are designed for these bolts) to ensure an even load. The results were disappointing, the padeyes distorting quite badly at 1.3 tons, and the bolts breaking at 1.7 tons.
>
> Clearly these padeyes have a number of useful functions aboard yachts, but they should not be used for harness attachment.

Panel 4
This panel was a balsa sandwich one with the U-bolts attached as per Panel 3 with 'penny' washers. At 1 ton pull the panel started creaking but it got to a peak of 2.25 tons before the panel failed in a similar way to Panel 3 ripping along the line of the bolt holes. The U-bolt showed signs of distortion, but the penny washers were badly dished.

Panel 5
This solid GRP panel had the U-bolt attached with a generous 7½in x 6in x ½in marine plywood backing plate, and 1 3/8in (35mm) penny washers. The normal washers were fitted between the penny washers and nuts.

The panel was strengthened by the plywood backing, but under 2 tons load distortion of the glassfibre starting splitting the plywood and, quickly afterwards, the panel itself, releasing the now slightly elongated U-bolt at 2.25 tons. The GRP panel split completely in two. The plywood backing plate allowed a good degree of shock absorption, and improved the installation considerably over Panel 3 although it split into three pieces.

Panel 6
This was a balsa sandwich panel with the U-bolt attached in an identical fashion to Panel 5 (ie plywood backing with penny washers). The panel started creaking at 0.8 ton, and showing sings of delamination at 1 ton. At 3 tons, with the panel badly distorted but still holding, the stainless bridge plate of the U-bolt started compressive bowing indicating severe elongation of the U-bolt. The panel failed at 3.25 tons when the plywood split completely in two, but could still hold static at 1.7 tons after failure.

Panel 7
This solid GRP laminate panel had the U-bolt fitted with a 6in x 3in (205mm x 100mm) aluminium backing plate, 3/16in (5mm) thick. The U-bolt was fitted with the normal washers.

The panel, although slightly reinforced by the aluminium, started bowing at 0.75 ton, the aluminium also showing signs of bending. The pull it provided was firm until 3.5 tons when the panel split, at 90 degrees to the line of the bolts. Even with the panel cracked across, it still withheld 1.4 tons.

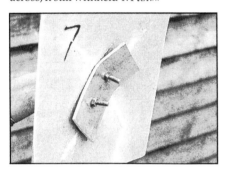

Panel 8
This sandwich panel had the U-bolt attached as per Panel 7 (ie aluminium backing plate). At 1 ton load there were few signs of any bowing, and at 1.5 tons it started creaking. At 2 tons the panel was becoming badly delaminated and distorted but still holding steadily. It withheld a 4 ton load before cracking up and releasing the load, although even when pull was re-applied it held 3 tons. The U-bolt was severely distorted with the bridge plate bowed.

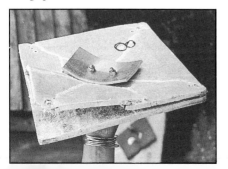

Conclusions
The graph (right) shows the relative strengths of fitting methods better than can be explained. The shock-absorption of the sandwich panels, and indeed the panels with plywood backing, is a significant advantage in a real-life situation in diminishing the maximum load of the crew member's fall by braking it. Panels 1, 2 and, to a lesser degree, 3, all showed disarming tendency to 'pop' the bolts abruptly.

Other than elongation, the U-bolts themselves showed no signs of failing, and although we didn't make tests, it is fair to assume that ¼in (6mm) stainless steel U-bolts could be employed without them becoming the weak link of the system.

We didn't test stainless steel backing plates because we felt this was overkill. In hindsight, and seeing how the aluminium bent allowing the breakdown of the panels, it would have been interesting to do so. It is our feeling (no more substantiated than that) that a stainless steel backing plate of the same dimensions would have yielded at least 20 per cent greater strength.

Solid panels
The use of penny washers instead of normal washers only increased the strength of the installation by 8 per cent.

A plywood backing plate, although inferior to aluminium because it split, provided excellent shock absorption through a combination of bending and drawing the washers into the laminations. We were slightly disappointed that it only yielded a 35 per cent increase in strength over a basic installation with ordinary washers (although that 35 per cent margin would more than likely be a lifesaver).

Clearly, employing a generous aluminium backing plate is an excellent method of attachment, offering 120 per cent increase in strength over basic washers.

Sandwich panels
Penny washers increased attachment strength by 25 per cent, significant compared to solid panel performances.

Although the sandwich panels started delaminating and breaking up quite early, this had to be a product, to some extent, of their small size and we believe they would hold longer in the larger panels you would find in a yacht (or in a small area such as the bridgedeck bulkhead in the cockpit their edges would be supported by panels adjoining at 90 degrees).

Fitting the U-bolt with an aluminium backing plate increased the strength over small washers by a very similar figure, 120 per cent, to solid panels.

Certain points arise out of our tests. First, that, whilst a *properly fitted* U-bolt is unlikely to present a weak link, if more than one person has his harness attached to it, there is a reasonable chance that the fitting will pull out in the event of both being thrown overboard in a knockdown.

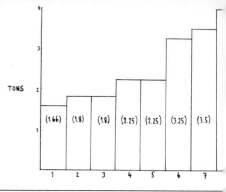

Relative strength of U-bolt attachment

Therefore strongpoints should be fitted for each crewmember it is anticipated will be in the cockpit at any one time.

Secondly, the standard padeye we tested was woefully inadequate for the job of harness attachment, for all its reasonably substantial looks.

The carbineer hook we tested failed at a figure greater than the BS4224 standard quotes for the breaking strain of tethers. The Gibb hook exceeded that, but is superior in design with its locking mechanism.

It seems that one of the most influential criteria of harnesses in use is the extent of energy absorption by the nature of the fall, plus the shock-absorption of the harness itself, the tether, attachment point to the yacht and indeed the human frame itself. Somebody falling overboard over the lifelines, but with his harness attached on the deck or in the cockpit, would have his fall cushioned by flexibility of the lifelines as the tether tensioned over them. In a similar way, glancing off a bulkhead or even tripping over the lifelines would absorb a good proportion of the kinetic energy compared to the abrupt freefall of the British Standards tests.

One area of concern is quite what these strains do to the body. A U-bolt might withstand 3 tons, but what about the rib cage, or does the survivor come aboard a vegetable? We were unable to track down any research conducted on this subject specifically, but believe it is significant that in the 1979 Fastnet there were no injuries caused by harnesses other than broken ribs.

One element of the tests that practicality didn't allow us to trial was the effects on U-bolts of being pulled from various directions. Our pulls were straight ones and it was assumed that the U-bolts were in the most common location aboard yachts – on the companionway bulkhead or bridgedeck bulkhead, or cockpit sole.

To a large extent the efficiency of a harness relies on the user employing it properly. The principal aim of a harness is to keep the person on board, rather than retain him once he's over the side. To this end, not only must it be attached to an established strongpoint, but it must be attached on the weather side to avoid the possibility of being thrown 12ft on a 6ft tether. Nobody should ever clip himself anywhere to leeward of amidships, and for safety's sake never clip to a point that already has someone attached to it.

Strength of tethers
The BS4224 standards stipulates that the safety line should have a minimum breaking load of 2,080kg (2.045 tons). A 10mm nylon (for shock-absorbing elasticity) provides this, but it is very important to note that *any rope's strength will be cut by 50 per cent if there is a knot in it*. A splice will cut down the breaking strain by 10 per cent, so tethers with splices at either end really need to use 12mm nylon.

Liferafts – the lessons learned

After the disaster searching questions were asked about yacht design and stability. But the questions didn't stop there. Crews who sought safety in liferafts had a rude awakening. This last line of defence short of swimming proved to have its own weaknesses

DESPITE THE weather conditions which are beyond understanding from the sidelines, of the liferafts carried, only fifteen were actually deployed. Seven lives were lost in incidents where crew abandoned ship in favour of the security of a liferaft, only for those vessels to be found subsequently afloat and towed to harbour. In comparison, fourteen lives were saved where liferafts were deployed from yachts which subsequently foundered. Liferafts were also used to transfer crew from yacht to helicopter. The longest any survivor remained in a liferaft was eight hours.

Those survivors who used liferafts were asked many questions and the Inquiry had a hard task drawing conclusions from the jumble of facts that emerged. But with liferafts capsizing frequently, some sustaining major damage, it was clear that design improvements were necessary.

Few crew had many problems in launching liferafts and, although some inflated upside-down, it took little time to turn the raft the right way up. Boarding the liferaft did prove a problem, however, with the access being on the opposite side to the painter attaching it to the yacht. Once in, the crew found they couldn't reach the painter to cut themselves free.

Fastnet report conclusions

3.75 Liferafts clearly failed to provide the safe refuge which many crews expected. Seven lives were lost in incidents associated with rafts, of which three were directly attributable to the failure of the raft and the yachts which these seven people abandoned were subsequently found afloat and towed to harbour. However fourteen lives were saved in incidents in which survivors took to rafts from yachts which were not recovered. Many crews used rafts successfully to transfer from yachts to helicopters or other vessels. It is asking a great deal of any very small craft to expect it to provide safe refuge in conditions which overwhelm a large yacht but this is what liferafts are expected to do.

Anatomy of a modern liferaft

The results of the research following the Fastnet Inquiry are immediately apparent in the modern liferaft. Large water ballast pockets all round the perimeter of the raft, weighted so they drop quickly into use, two inflation tubes, either capable of supporting the liferaft, a self-inflating arch supporting the canopy at right angles to the attachment point of the cone shaped drogue trailed on the end of a very long line.

The inflatable floor is not for comfort. The greatest danger once aboard the raft is exposure, especially for crew who are cold and wet. Body heat is lost through the floor of the raft and the inflatable floor provides an essential barrier against this silent but deadly enemy. Other features include safety lines all round the outside and inside of the raft. Crew members were washed out of liferafts in the Fastnet.

Access to the raft is via a suitably wide opening similar in design to a climbing tent entrance with a simple velcro/tape closure. The entrance has to combine width with security. A large-framed man complete with lifejacket has a significant girth. Some crew members in the Fastnet had problems getting into their raft with one entering via the observation port. But the contrary problem of a too wide and insecure an opening allowing water to wash into the raft also has to be avoided.

Another aid is emergency lighting both inside and outside the raft.

The amount of kit inside the raft will vary greatly from model to model, the most important item being the sea drogue which has to be streamed quickly.

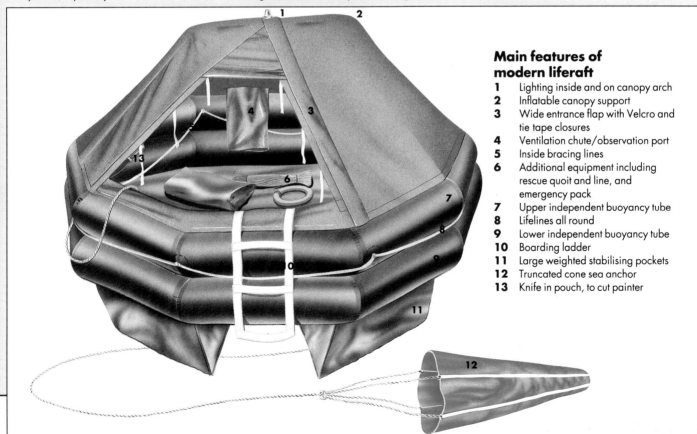

Main features of modern liferaft

1. Lighting inside and on canopy arch
2. Inflatable canopy support
3. Wide entrance flap with Velcro and tie tape closures
4. Ventilation chute/observation port
5. Inside bracing lines
6. Additional equipment including rescue quoit and line, and emergency pack
7. Upper independent buoyancy tube
8. Lifelines all round
9. Lower independent buoyancy tube
10. Boarding ladder
11. Large weighted stabilising pockets
12. Truncated cone sea anchor
13. Knife in pouch, to cut painter

Many other detailed points came out such as the necessity to have grab lines all around the outside and inside of the raft. A yacht 'rescued' a liferaft but was unable to secure it or gain access to the canopy entrance. The raft drifted away.

Two major areas of concern arose. The first was the structural integrity of the liferafts as some disintegrated with canopies coming adrift and flotation tubes separating. Second was the propensity of liferafts to capsize.

It is this second failure that led to major research work being carried out under the auspices of the Department of Trade and Industry linking with research also being conducted by the Icelandic Government. As a result of full-scale trials carried out in Force 14 gales off Iceland in February 1980 and work by the National Maritime Institute, recommendations were made that have had a far-reaching effect on current design practice.

Before the sea trials, it was assumed that a liferaft obtained its stability from the water ballast pockets underneath the flotation tubes. The trials showed conclusively that the sea anchor was a major factor. Those liferafts that lost their drogues capsized. The

Some of the changes made by Avon on four-man raft

	Old	New
No of ballast pockets	4	5
Total water capacity	100 lb	275 lb
Weighted	no	yes
Approx time to fill	120sec	45sec
Pocket material (oz/sq yd)	4¼	7
Sea anchor	square	cone
Length of line (metres)	10	20

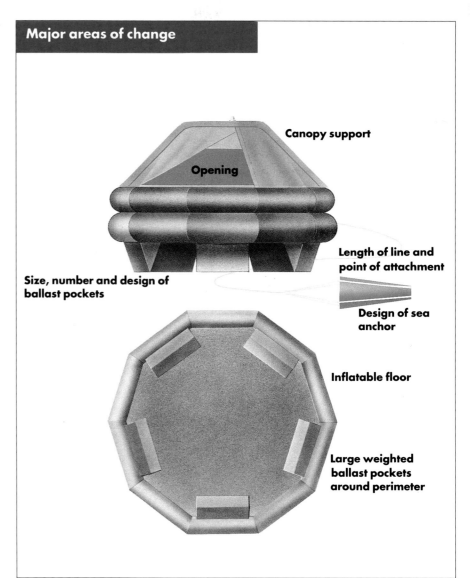

Major areas of change

- Canopy support
- Opening
- Size, number and design of ballast pockets
- Length of line and point of attachment
- Design of sea anchor
- Inflatable floor
- Large weighted ballast pockets around perimeter

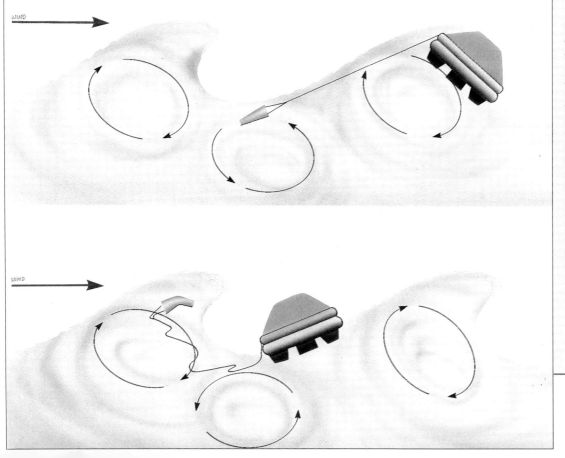

It is easy to see how enormous snatch loads can be generated on the sea anchor and its line. At one point the underlying motion of the water has the liferaft driving one way with the sea anchor being driven in the opposite direction. If the sea anchor then gets caught up by a breaking wave, it can be tripped and may get tangled. When it falls into the next trough, it will have the full force of the liferaft being driven by both wind and water to contend with

standard design of the drogue was also found to be at fault becoming tangled or failing completely, some separating from the raft.

Subsequent research showed that not only was the design of the drogue important, but that both the length and strength of line was critical. It was discovered that in only 20 knots of wind and 1/1.5 metre waves, the snatch loads were reaching ten times the normal loading due to the raft accelerating down one wave whilst the drogue was being driven in the opposite direction by the circular motion of the water in the preceding trough.

Once deployed, it was found that the entrance to the raft should be on the opposite side to the attachment point of the drogue line. Otherwise the entrance faced directly into the wind, the incoming draught causing the raft canopy to 'balloon' and lose stability.

Ballast pockets were still found to be critical to stability but to function in those vital first few seconds after launch they needed to be large, sited as near the outer edge of the bottom flotation tube as possible and weighted so they dropped into position and filled quickly.

The aspect that the supporting canopy presented to the wind was found to be of importance. With circular liferafts, the self-inflating canopy support should be at right angles to the wind direction.

Even the number of crew in the raft and where they sat was found to affect stability. Too few crew for the size of raft or crew sitting to leeward side aided capsizing.

Liferaft standards

There are a variety of specifications for liferafts but no one set standard. Liferafts constructed to intergovernmental SOLAS (Safety Of Life At Sea) standards, which apply to all merchant vessels and passenger ships, should, among other features, be capable of surviving in any weather for at least 30 days, have at least two buoyancy compartments, either one capable of supporting the raft, and an inflatable floor and two layer canopy, both aimed at protecting the crew from exposure in cold waters. As they are intended for longer use they contain an extensive pack of survival equipment. They are also very expensive.

In the UK, offshore yachts over 13.7 metres are required to have sufficient liferafts for all crew members to Department of Transport Marine Directorate standards. In practice, this means near to SOLAS standards, the principal difference being the nature of the pack enclosed with the raft.

Ordinary pleasure yachts under 13.7 metres are not required to carry liferafts but various bodies, particularly those involved with racing, set their own standards.

Liferafts may be constructed to meet RORC (Royal Ocean Racing Club) and ORC (Offshore Racing Council) requirements. Their standards are slightly less stringent than those set by SOLAS. There is no requirement, for example, that the liferaft be inspected during manufacture and the survival pack is much less extensive.

Other governments, quasi-governmental bodies and associations outside the UK have their own requirements.

Liferafts are also available built to their makers' own published standards. Some may be perfectly adequate for offshore use; others, as the makers make clear, are intended for use by coastal yachtsmen where there is a reasonable expectation of rescue within a relatively short period of time.

Some tenders, rigid or inflatable, suitably adapted, double as coastal liferafts, giving the crew the added option of being actively involved in their own rescue rather than relying on the emergency services.

Cost

Yachtsmen are notoriously keen to cut costs, especially on non-essential equipment. Whether you put a liferaft in this category depends on your own outlook. A lot will no doubt depend on the amount of offshore sailing you do, whether you are involved in club racing, the size of yacht and income available.

Typical costs of four-man raft
SOLAS	£2,150
RORC (in canister)	£1,200
Coastal	£840

To this must be added an annual service charge which will depend largely on how well you look after the raft. The annual inspection and repacking fee by an approved agent will be in the order of £50-£70. If the raft is stored so that sea water/sunlight can penetrate, or the canister is used as a step, then expect much higher costs as repairs will be needed.
An alternative is to hire a raft, either for the whole season or for the summer cruise.

Typical costs four-man liferaft hire
SOLAS	(annual)	£437
RORC	(annual)	£275
RORC	(month)	£138
RORC	(week)	£43

Conclusions

The loss of life that led to the Fastnet Inquiry report ultimately led to a reappraisal of liferaft design. We have all benefited. No doubt some yachtsmen now owe their lives to the improvements that were made. But no help can be given over the decision on whether to abandon ship. That will always remain, as it did in August 1979, the decision of the skipper and crew, based on how they see the situation.

Stowage – Above or below?

Above

Advantages	Immediate availability
	Easy launching
Disadvantages	Danger of being washed overboard by breaking seas
	Continuously exposed to marine environment leading to possible unseen damage due to leaking seal round canister
	Needs additional protective canister adding to weight and bulk
	Obstruction on deck
	Ease of theft

Below

Advantages	Protected environment
	Out of way
	Heavy weight located low down
	Stored in valise
Disadvantages	Physical effort involved in bringing on deck quickly particularly in heavy seas
	Isolation of raft if fire below or serious flooding
	If yacht rolled, chaos below can obstruct passage of raft through to companionway
	Time taken to launch

Current design solution is to locate raft in/around cockpit, under floor in special locker or aft for quick release over stern. This combines an element of protection with ease of launching. Wherever located, must be both secure in event of yacht rolling, being washed by breaking seas and yet be easy to release.

Careful stowage of the liferaft is essential. If on deck, it should be secure but capable of being released quickly. If below in a soft valise, it must be accessible

Stowage

3.60 Twelve liferafts were washed overboard, of which eight were stowed in the cockpit and four on deck. In several cases, rafts were secured in place only by the lid of a locker, so that as soon as the locker was opened, which some did accidentally, the raft either fell or was washed out. One of the deck stowed rafts which was lost went overboard still secured to the chocks on which it was stowed.

- **80 per cent of current production yachts have no dedicated harness eye**
- **30 per cent of yachts have insufficient seaberths**
- **60 per cent of cookers are not properly secured**
- **90 per cent of companionways are not storm proof**

Survey results

The Fastnet storm exposed many weaknesses in the yachts comprising the fleet. Taking the recommendations of the Fastnet Report, the ORC Special Regulations Category 2 and our own experiences, we surveyed over fifty current production yachts and found that in standard trim few production yachts could be described as 'designed and equipped for offshore sailing'

THIS SURVEY was not intended to reveal basic structural deficiencies or unsound stability profiles. Its purpose was to highlight the details which make an otherwise sound yacht either safe and comfortable or potentially hazardous and downright awkward.

Jackstays and harness eyes. Keeping the crew on board during bad weather is a fundamental safety requirement. Jackstays and strong points on which to attach safety harnesses, lifelines and pulpits are the first and last lines of defence.

Yet only one boat surveyed had jackstays fitted, as standard while only 12 per cent had a harness eye positioned where a crew could hook on before coming on deck as required by the ORC. Only 20 per cent of boats had eyes within reach of the helmsman in his normal steering position. Most modern yachts have guardrails and pulpits which conform to good practice.

All boats should have one strong-point near the companionway and one on each side of the cockpit within reach of the helmsman. The jackstays should form a clear run from bow to stern and be of uncoated stainless steel wire. The terminals should be swaged bottlescrews. While the jackstays themselves are easy enough to fit after taking delivery, the deck eyes are best put in during building.

Companionways. During the Fastnet storm companionways were the main source of flood water getting below. Forty-two per cent of yachts reported significant water entering through the companionway. Many skippers kept the companionway open because they were afraid crew would be trapped below if the hatch was shut. As a result, the report strongly recommended that hatches should be designed so that they could be opened and closed from above or below decks.

The report also noted the number of boards which fell from yachts during capsize. It is now a requirement of the ORC that hatches be securable from above and below decks and that washboards be attached to the yacht.

Our survey revealed that under 10 per cent of yachts had companionway openings which conformed to this requirement and this figure was swelled by three boats with hinged opening doors. Although it would be simplest and best for this sort of lock to be fitted as standard, proprietary locks are available, notably the Fastnet from Sea Sure.

Even fewer yachts had hatchboards which could be secured – just 2 per cent though this is quite a simple owner's modification requiring a length of line and a couple of jam cleats.

A bridgedeck of sufficient height is another element in keeping water out of the interior. The higher this is the better. The ORC stipulates that the bridgedeck should reach maindeck level or be capable of being sealed off to that level. In practical terms, a bridgedeck to the level of the cockpit seating is sufficient. Yet even this standard is not met by 30 per cent of the production yachts surveyed.

Cockpit lockers. In normal conditions no one would think twice about cockpit lockers as a point of weakness. But in extreme conditions they are very vulnerable, particularly the full depth variety. Water getting in here goes straight to the bilge. There is a conflict, though, between the need for locker tops which allow easy access for large items such as dinghies and sails, and security. The most secure yacht in this respect is one without cockpit lockers. But in practice at least one and often two full depth-lockers are found.

What the Fastnet Inquiry said

Watertight integrity. The most serious defect affecting watertight integrity was the design and construction of the main companionway. It is recommended that the Special Regulations relating to the blocking arrangements of the main companionway should be extended to introduce specific requirements for the blocking arrangements to be totally secure but openable from above and below decks... It is also recommended that the Special Regulation relating to bilge pumping should require bilge pumps to discharge overboard and not into the cockpit unless the cockpit is open to the sea.

Comfort and security of accommodation. It is evident that the stowage arrangements in some boats are designed to be effective only up to 90 degrees of heel. It is recommended that the Memorandum on Safety should draw attention to the need for the securing arrangements for heavy items or equipment and all stowage to be effective in the event of a total inversion.

Safety harnesses. In spite of an adequate Special Regulation and a paragraph in the Memorandum on Safety, six lives are believed to have been lost through the failure of safety harnesses or their attachment points. It is recommended that the RYA and the RORC should draw attention to the importance of the following points:

1. The need for harnesses which comply to BS4224, which are regularly surveyed and maintained and for which strong attachment points are available.

2. The need for double harness lifelines in severe weather conditions.

3. The danger of clipping on to guardrails, as in heavy weather these do not necessarily constitute strong attachment points.

4. The need for an adequate deck line or lines led from cockpit to a point forward of the mast for use as a harness attachment point, and the advantages of having permanent life lines in suitable places which can be clipped to harnesses.

In addition we would like to emphasise the practical advantages of a harness which is manufactured as a combination harness and lifejacket.

Engines/electrics. Several damaged yachts retired safely under power. There is also some evidence that the use of engines improved the manoeuvrability of yachts picking up survivors and in some cases assisting in maintaining steerage way in storm conditions. In addition, the use of engines for maintaining battery power was shown to be of importance. The RORC should consider whether engines should not be mandatory for safety reasons and whether alternative methods of starting engines should be required when the starting battery is flat.

A harness eye should be placed close to the companionway so that crew can clip on before coming on deck. Survey rating 12 per cent

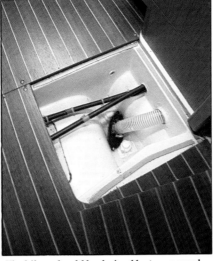

The bilges should be drained by two manual pumps, one operable from above and one below decks. You must be able to get at the topside pump without opening lockers or hatches. Survey rating 12 per cent

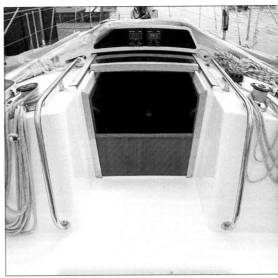

Companionways should be parallel sided, the washboards should be capable of being fixed in position and the hatch opened from above or below deck. Survey rating under 10 per cent

Most gas bottles are properly stowed in their own compartment with overboard drains but too few are sufficiently well secured against a knockdown. Survey rating 31 per cent

All hull openings should be fitted with proper seacocks and the pipework should be double clipped. Survey rating 72 per cent

Cookers must be positively held in their gimbals, the crash bar must be stout and a safety strap should be available to hold the cook in position. Survey rating 40 per cent

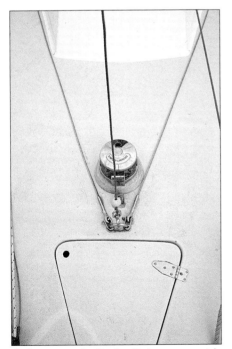

Jackstays must be fitted so that crew can remain clipped on while moving from astern to the bows. Lines along the sidedeck are good, lines along the coachroof are probably better, provided the coachroof extends far enough forward. Survey rating under 2 per cent

Small tops which do not break the line of the coaming, and which are seated on to a seal with a draining gutter round, are best. Three hinges and two catches are needed on all but the smallest lids.

This is how production yachts stacked up against this standard. Six per cent had no cockpit lockers, 25 per cent had one and 60 per cent had two. Nine per cent had more. No yacht had effective seals round the locker but all had gullies of reasonable depth to drain water away. Of hinges, 85 per cent had two hinges and 10 per cent had a single, full length hinge. Ten per cent of yachts had no catches at all, 70 per cent had one and 20 per cent had two. We were encouraged that 75 per cent of yachts had lockers which did not break the line of the coaming.

Bilge pumps. Before leaving the subject of water getting below, the ORC demands that two bilge pumps be fitted, one of which can be operated from above and one from below

This excellent battery box is better than most. They must be held firmly, protected from bilge water and vented to remove dangerous gases. Survey result 88 per cent

decks and neither needing hatches or locker tops to be opened. In practice, 70 per cent of yachts had a single pump and just 12 per cent two manual pumps installed correctly. However, 24 per cent had an auxiliary electrical pump which could be operated from below.

Seacocks. The proper fitting of seacocks to all skin fittings (except cockpit drains and scuppers) has long been a point of concern. However, we are pleased to be able to report that all boats surveyed had adequate seacocks installed and that 72 per cent had them double clipped as required by the ORC.

Below decks

One of the conclusions of the Fastnet report was that stowage arrangements on yachts were only effective up to 90 degrees. In the event of B1 and B2 knockdowns, all sorts of gear from light items of food to cookers, batteries and bottles began flying around. When we checked to see if manufacturers had taken heed of this caution, we found a depressing situation.

Cookers. Sixty-three per cent of cookers would come clear of their gimbals at more than 90 degrees.

Heavy gear. Only 26 per cent of boats had positive stowage for heavy items of equipment though a further 30 per cent had some sort of stowage which could be made suitable.

Batteries. Twelve per cent of boats had batteries insufficiently well-secured.

Gas bottles. Thirty-one per cent of gas bottles would not remain in their stowage in the event of a knockdown.

Leecloths. It remains a mystery why builders do not fit leecloths. They are simple, quick and cheap, and anyone sailing offshore at night will need them. Of the Fastnet Race fleet, only 8 per cent had insufficient seaberths (with leecloths) for half the crew, yet in the showrooms only 10 per cent had them fitted. Just over 50 per cent had suitable bunks without leecloths and 33 per cent could not provide suitable seaberths for half the crew with or without cloths.

In our terms, a suitable seaberth is one which is substantially straight and on the fore and aft line, 6ft 2in or more long, not wider than 24in and situated aft of the mast.

The galley. The person possibly most at risk below at sea is the cook. With hot liquids, heavy pans and a swinging work surface to contend with, the opportunities for accidents are many. All boats are fitted with gimballed cookers these days, though many cannot swing sufficiently without hitting the hull-side.

Most cookers are fitted with fiddle rails too, but nearly half are not provided with a crash bar to prevent the cook being thrown against the cooker or to provide him with something to grab hold of. Equally important is that he must be able to retain his balance while working with both hands, and for this a restraining strap is needed. No more than 16 per cent of boats are so fitted.

Engine starting. One of the aspects of seamanship which is sometimes overlooked and which we have been emphasising is the reliability of the auxiliary engine. But an engine is only of use if it can be started. Battery failure is not uncommon so back-up systems are required. Our survey found that, out of all yachts, only 10 per cent could be started by hand either because of the size of the engine or the way it was installed. On the other hand, 78 per cent had two or more batteries.

But this does not tell the whole story. It is unlikely that anyone will be able to start an engine of three cylinders or more though it has been done. For these a battery system of two or more is essential. One of these batteries should be dedicated to engine starting, but the others should be capable of being linked up in an emergency.

Of three-cylinder engined boats, all those surveyed had twin-battery installations.

The last word

Our survey confirmed our feeling that modern yachts, while looking very pretty and being in many ways ingenious, lack many of the essential requirements for offshore sailing.

All the things we looked at could be dealt with cheaply and easily by the builder during the construction process. None of them would be detrimental in any way to looks, harbour efficiency or accommodation plan.

The market place is competitive and there is great pressure on companies to reduce the 'bottom line' price as far as possible. The result is that more and more essential equipment gets squeezed off the standard and on to the optional list. Experienced yachtsmen can spot this and make allowances. First-time buyers cannot.

But it should not be necessary for any buyer to have to tot up the cost of essential safety. A boat sold for cruising should be fitted out for cruising within the standard price.

When it comes to loose gear such as flares, liferafts, lifebuoys, etc, where there is considerable brand choice and many levels of expenditure, it is reasonable to leave the selection to the customer. But fitted equipment or design features, where there are few if any options should be part of the basic package. Manufacturers must accept this responsibility. ☐

Right: the Yachting Monthly survey revealed that very few modern production boats in standard form comply fully with ORC Special Regulations for offshore sailing

---

What the ORC regulations (Category 2) say

- When an electric starter is the only provision for starting the engine, a separate battery shall be carried, the primary purpose of which is to start the engine.
- Ballast and heavy equipment. All heavy items including inside ballast and internal fittings (such as batteries, stoves, gas bottles, tanks, engines, outboard motors etc) shall be securely fastened so as to remain in position should the yacht capsize 180 degrees.
- Yacht equipment and fittings shall be securely fastened.
- The main companionway shall be fitted with a strong securing arrangement which shall be operable from above and below.
- All blocking arrangements (washboards, hatchboards, etc) shall be capable of being secured in position with the hatch open or shut and shall be secured to the yacht by lanyard or other mechanical means to prevent their being lost overboard
- Cockpit companionways, if extended below main deck level, must be capable of being blocked off to the level of the main deck at the sheerline abreast the opening. When such blocking arrangements are in place, the companionway (or hatch) shall continue to give access to the interior of the hull.
- Cockpits shall be structurally strong, self-draining and permanently incorporated as an integral part of the hull. They must be essentially watertight, that is, all openings to the hull must be capable of being strongly and rigidly secured. Any bow, lateral, central or stern well will be considered as a cockpit.
- Seacocks or valves shall be fitted on all through-hull openings below LWL, except integral deck scuppers, shaft log, speed indicators, depth finders and the like; however, a means of closing such openings, when necessary to do so shall be provided.
- Soft wood plugs, tapered and of the correct size, shall be attached to, or adjacent to, the appropriate fitting.
- Lifeline terminals and lif-line material. Where wire lifelines are required they shall be multi-stranded steel wire. A taut lanyard of synthetic rope may be used to secure lifelines provided that when in position, its length does not exceed 4in (100mm).
- Jackstays. Wire jackstays must be fitted on deck, port and starboard of the yacht's centreline to provide secure attachments for safety harnesses. Jackstays must be attached to through-bolted or welded deck plates, or other suitable and strong anchorages. The jackstay must, if possible, be fitted in such a way that a crew member, when clipped on, can move from a cockpit to the forward and to the aft end of the main deck without unclipping his harness. If the deck layout renders this impossible, additional lifelines must be fitted so that a crew member can move as described with the minimum of clipping operations.
- A crew member must be able to clip on before coming on deck, unclip after going below, and remain clipped on while moving laterally across the yacht on the foredeck, the afterdeck, and amidships. If necessary additional jackstays and/or through-bolted or welded anchorage points must be provided for this purpose.
- Through-bolted or welded anchorage points, or other suitable and strong anchorage, for safety harnesses must be provided adjacent to stations such as the helm, sheet winches and masts, where crew members work for long periods. Jackstays should be sited in such a way that the safety harness lanyard can be kept as short as possible.
- The cooking stove must be securely installed against a capsize with safe, accessible fuel shut-off control capable of being safely operated in a seaway.
- Bilge pumps. At least two manually operated, shall be securely fitted to the yacht's structure, one operable above, the other below deck. Each pump shall be operable with all cockpit seats, hatches and companionways shut.
- Unless permanently fitted, each bilge pump handle shall be provided with a lanyard or catch or similar device to prevent accidental loss.

Chapter 9

Electrical systems

From the basics such as navigation lights to the convenience of instrument and navigation systems and the comfort of heating and refrigeration, electrical power is central to the everyday running of today's yachts. And yet electrical failure is one of the most common problems afloat. In itself this might not jeopardise the safety of the yacht, but it will certainly affect her efficiency and so ultimately her seaworthiness

THE EXTRAORDINARY growth in the electrical requirements of modern boats has exposed the inadequacy of older yachts in this area. Even modern boats are often far from amply equipped.

Electrical installations on boats are improving all the time. They need to. Not only were they once generally pretty dire, the strains placed on the electrical fabric of a modern yacht is constantly increasing.

The system on any yacht, however simple or sophisticated, breaks down into four areas: creation, storage, supply and end use. All are inextricably interlinked and depend ultimately on the overall electrical requirements of the boat, the type of equipment fitted and the amount it will be used.

The first important point to realise is that however good a yacht's original electrical installation, it is working against several disadvantages. The first and most destructive is the marine environment: salt, water and electricity are unhappy bedfellows. The second is motion and vibration which lead to chafe and broken and loose connections and the third is the yachtsman's penchant for adding new electrical and electronic equipment with scant regard for the ability of the original installation (alternator, battery and wiring) to deliver.

The first and second are overcome by including the electrical installation on your maintenance list – instead of waiting until a component stops working (like engines, it will always be when under maximum strain and when most needed). And the best philosophy to adopt when adding new equipment is to think of your battery as a pump with limited capacity. If, for example, it was a central heating pump, would you expect to be able to add three new radiators and still be able to deliver piping hot water to the whole system? Would the boiler (alternator) have the power to deal with the increased water capacity in the system and not deliver luke warm water around the house?

Geoff Pack in his article *Watt price amps?* (July 1989) went into the business of working out how much electricity your boat consumes and how much must therefore be supplied, so this will not be dealt with here. This time, we are looking at the actual fabric of the system – the wiring, batteries, alternator and so on and how to make them capable of coping with the rigours of extensive sailing offshore.

1979 Fastnet Race Enquiry Report

- Several yachts reported losing the use of all electrics or one or more items of electrical equipment due to flooding
- 77% of the fleet used normal navigation lights throughout the storm and 68% were aware of the presence of other yachts [from their lights] in their vicinity at the height of the storm
- 16% of the fleet reported major difficulties with compass or cabin lighting
- The figures below are an estimate of normal battery capacity crews estimated they had available during the storm

  Below 25% capacity 9%
  25% – 50% capacity 10%
  51% – 75% capacity 23%
  75% plus capacity 47%

Creation

On most boats this is the engine driven alternator with an output of between 35 and 55amps. This is perfectly adequate for most medium-sized yachts with the usual array of lights and electrical equipment. A useful rule of thumb is that the alternator capacity/output should be at least 25 per cent of the total battery capacity. However, as soon as you start adding such equipment as a refrigerator, electric windlass, central heating or extra electronic equipment this output may become insufficient.

It is tempting to think that merely adding batteries will solve the problem of power supply. This is only the case when you have an alternator to suit and to balance the creation and storage of electrical power.

The way to calculate the charging rate required by your system is to add up the maximum continuous load which will be put on the battery at any one time, multiply that by a small factor to take into account inefficiencies in the battery itself and add the required battery charging rate.

For example, during a night passage the *maximum* electrical load may well include 3amps for the sidelights, 4amps for the VHF (transmit), 3amps for cabin lights and a further 3amps for ancillaries such as Decca and instrumentation making the total load in the region of 13amps. Multiplying this by, say 1.4, to allow for inefficiencies gives a figure of 18.2amps. Since we will also be wanting to charge the battery at a reasonable rate (usually what is known as the 10 hour rate), we must allow say 12amps for a 12v battery giving a total charging requirement of 30.2amps, well within the 'normal' range of standard alternators. But as soon as extras such as a fridge (2amps), heater (2.3amps), autopilot (0.75amps), radar (5amps) and water pump (3.25amps) are added this figure comes out at just below 50amps suggesting that a larger alternator might be needed.

If you decide to change your alternator, make sure it is one which works at a speed

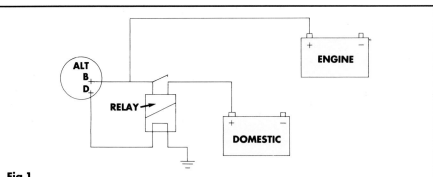

Fig 1
Split charge relay. The electrically operated relay parallels the alternator output to each battery or batteries only when the alternator is running. When the engine stops the relay falls open

NB: A battery isolating switch should still be fitted

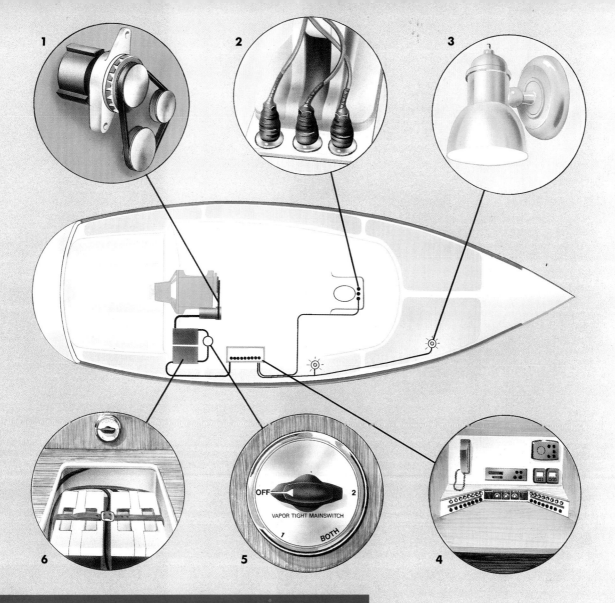

Essential elements

1
Electricity creation. The output of the alternator must match the demands of the electrical system and the capacity of the batteries. As a useful rule of thumb the alternator capacity/output should be at least 25 per cent of the capacity of the batteries. If extra equipment is fitted merely adding batteries is not the complete solution. The alternator may also need changing to one with higher output

2
Deck plugs, the most troublesome connection of them all. Use waterproof plugs of the Bowdeck type, clean terminals and smear them with silicone grease and for final protection wrap them well with self-amalgamating tape. The best solution, though more of a problem when it comes to taking the mast out, is to lead the wires through the deck via deck glands and make the connection below decks

3
Even below decks appliances are subject to damp and the salty atmosphere. Connections are again the problem though a wax spray such as Waxoyl give good protection

4
The switch board is the key distribution point for all electrical systems on board. Wires behind should be left long enough to allow the front panel to be removed easily and each circuit should be separately switched and fused – better still using circuit breakers. If additional heavy drain equipment is added (eg fridge, heater, radar) the supply wires to the panel may need uprating

5
A good quality isolator/battery switch is essential as poor connections here can lead to substantial voltage drop both when charging and using power. It is vital to have a switch that 'makes' before it 'breaks' to avoid alternator damage when switching between batteries when charging. As a general rule it is best to charge to individual batteries rather than having the switch in the 'BOTH' position for efficient and fast charging

6
Batteries must be securely fastened down with some kind of strap and firmly wedged in the box to prevent movement. The box should be well ventilated to allow battery gases to escape. Battery tops and terminals must be kept clean and dry to prevent current leakage

compatible with the engine. A slow running diesel may never spin a high speed alternator sufficiently fast for it to produce the required current. This is not only inefficient, it is potentially damaging to the alternator since its cooling vanes will not be doing a good job. The answer may be to fit a larger pulley to the flywheel but bear in mind that the water impeller might be driven off the same belt and its rotating speed will also have to be considered. The wiring supplied with the engine will be adequate for the standard alternator but if you move up a grade, the wiring will also have to be upgraded.

Charging. When installing a twin battery system or converting a single battery system, some thought should be given to the way the two batteries or banks of batteries will be charged.

A split charging system is a way to safeguard the dedicated engine start battery and keep it fully charged at all times. Two basic types are available. One is the split charge relay (**Fig 1**) which is an electrically operated relay switch that parallels the alternator output to each bank of batteries only when the alternator is working. When the engine stops the relay switches so isolating the batteries from one another and 'protecting' the engine start battery from discharge when the yacht's electrical services are in use. The main drawback of such a system is that you are relying on a mechanical component to perform faultlessly and in time, like the points of a car, the contacts will burn out.

The second option is a blocking diode (in effect an electrical one way valve) which has no moving parts to wear and which can be used in one of two ways (**Fig 2 and 3**). However, although more reliable than relays, blocking diodes suffer one drawback in that their in-built resistance causes voltage drop and hence reduces the ability of the alternator to charge the batteries.

The simplest and best method is to use a manually-operated four-way battery switch which directs the current from the alternator to the battery of your choice. The switch allows current to be drawn from either or both batteries and for the charging current to be similarly directed. The wiring for the switch is straight forward, but the key feature must be that it 'makes' before it 'breaks'. In other words, while switching from one battery to the other or to both, at least one battery is always in contact with the alternator. The reason for this is that if the alternator is open circuited it can damage its diodes.

If fitting or replacing a battery switch it pays to go for good quality. There are certain cheap makes of Far Eastern origin which deteriorate rapidly leading to voltage drop or even worse as some don't 'make and break' properly when being switched. The switch should move smoothly with well-defined clicks and if the one you've already got fitted seems loose or grates at all when moved then it's time for a replacement.

One of the problems with conventional battery charging is that the cells are seldom more than about 70 per cent charged. Add to this the fact that only about two-thirds of the batteries nominal capacity is available before the voltage falls too low to be usable, and it is clear that a 100ah battery is actually only capable of delivering a paltry 30ah or so. Nothing can be done about the lower figure but the degree to which the battery can be charged can be improved.

On most alternators a fixed voltage regulator is fitted as standard. This takes no account of any losses in the electrical system and consequently a good charging voltage

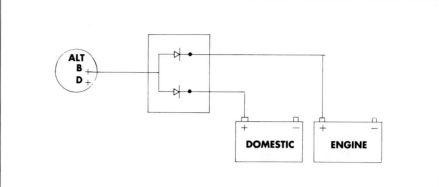

Fig 2
Alternator output to a blocking diode with two or three outputs. Each output is then connected to a separate battery or bank of batteries

NB: A battery isolating switch should still be fitted

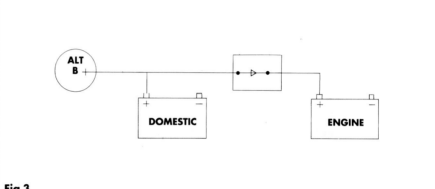

Fig 3
Alternator output is taken direct to the domestic battery with a feed to the engine battery. A single diode is fitted in the line to prevent the engine battery discharging into the domestic

of 14v is sadly diminished to around 13v or less due to losses through cable size, long cable runs, badly-made connections or faulty changeover/isolating switches. As a way of resolving this problem various types of variable output regulators have appeared on the market. Baron Instruments introduced a manually-adjusted little grey box and the Solent charger is another manually-operated variable output regulator. Perhaps the best known is the TWC electronic regulator. Using microchip technology this unit is able constantly to monitor true battery status and temperature at the battery terminal. It compensates for any losses between alternator and battery and applies exactly the right amount of charging voltage to the battery.

The TWC has been on the market for several years and in practice has proved remarkably effective. One report we have heard is from a yacht taking part in the 1987 Fastnet Race having to charge batteries every four hours. After fitting a TWC regulator it sailed virtually the same course the following year in the Triangle Race and needed only to charge for 50 minutes in 56 hours' sailing.

To ensure that batteries are as fully charged as possible it is important to realise that charging voltage is vital (as well as current). A 12v battery will never be fully charged with only 12v. Between 13 and 14.5v is required.

One of the most important checks that can be carried out is to measure output voltages at various points using a digital voltmeter. First of all this is measured at the alternator output, next (if fitted) at the diode output, then at the battery switch and finally at the battery terminal. The table below gives the sort of readings you could expect with a fixed voltage regulator compared to a TWC.

	Fixed voltage regulator	TWC
Alternator	14.22v	15.52v
Diode output	13.40v	14.70v
Battery switch (out)	13.30v	14.6v
Battery terminal	13.20v	14.5v

The reason for the differences is simply voltage drop – or resistance. The blocking

ORC Special Regulations, Category 2

We have chosen these regulations, which are safety and equipment requirements for offshore racing yachts, as a model, which is for yachts taking part in races 'of extended duration along or not far removed from the shoreline or in large unprotected bays or lakes where a high degree of self-sufficiency is required but with a reasonable probability that outside assistance could be called upon for aid in the event of serious emergency'.

Below is a précis of the requirements for electrical equipment with the paragraph numbers referring to those in the Regulations. Copies of the booklet are available from the **ORC**, 19 St James's Place, London SW1A 1NN (price £2.00, plus 20p p&p).

- When an electric starter is the only provision for starting the engine, a separate battery shall be carried, the primary purpose of which is to start the engine (5.1).

- All heavy equipment...such as batteries...shall be securely fastened so as to remain in position should the yacht be capsized 180° (5.4).

A well laid out switch panel and one that is easy to work on. There is enough slack in the cables to allow the panel to be completely removed

Above: chocolate box connectors of this type are quite adequate provided they are suitably protected with a wax spray to keep the moisture away. If the connection is one that is unlikely to need attention, the whole can be encapsulated in silicone sealant. Left: connections are the weak link in any electrical system. Wire ends should be soldered for a clean connection as even the odd one or two stray wires making no or poor contact can lead to voltage drop

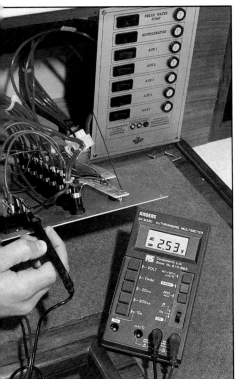

A digital multimeter is a vital tool for finding faults. It can also provide a very accurate and fascinating insight into the amount of power consumed by individual items of electrical equipment

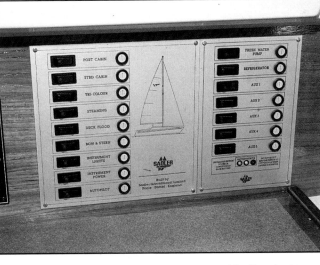

Many modern switchboards have a schematic diagram of the boat with small lights that indicate which navigation lights are in use. Photographs by Patrick Roach

also damage sensitive electronic equipment.

The size of batteries required depends on the amount of electrical and electronic equipment on board. The power requirements on modern yachts dictate that two or more batteries are highly desirable, one of which is dedicated to starting and one or more to the yacht's services.

For yachts up to 35ft or so a starting battery of 60 or 90ah capacity should be adequate with one of 100 to 120ah for the service side. Yachts with more than the usual round of lights, VHF, electronic navigator and instruments should consider having two service batteries of about 100ah each. When building up banks of batteries wired in parallel to increase capacity, be sure to use similar types. Batteries with different capacities, discharge and charge rates will be continually filling and emptying from each other shortening their life.

Although alkaline or NiCad batteries have much to recommend them, particularly for serious offshore sailing, for financial reasons, the lead acid battery is universally installed.

However, within this category are now a number of different types. The conventional sort have a liquid electrolyte which needs topping up at regular intervals, is susceptible to being tipped beyond a certain angle and gasses at a relatively low voltage.

'Low maintenance' or 'maintenance free' batteries can take a higher charge before gassing. Although they are not fitted with refilling plugs because the electrolyte is used up at a very slow rate, they are not hermetically sealed and should be checked annually and topped up if necessary. This sort of battery has calcium reinforced grids which makes it easier to recharge and so is

diode performs its task, but at no small cost, and the other losses are incurred by bad connections, cheap and cheerful switches and poor wiring. This test will quickly reveal any problems.

Storage

The batteries are the heart and soul of a good electrical system. The right size and type of batteries in good condition, properly charged will give long, reliable service. Inadequate, tired, half full ones will not only be continually letting you down but may

ideally suited to marine use. However, they do not like being deeply discharged and if they become completely flat they are permanently damaged. The voltage should not be allowed to fall below 10.5v.

An improved version of the maintenance free battery is the hermetically sealed type which has a number of advantages but is very expensive. From a marine point of view it is totally undamaged by being inverted and never needs topping up, any gasses created being turned back into water. It has good recharging properties but does not like

Electrical maintenance checklist

Batteries
- Check electrolyte levels and if necessary top up with distilled water
- Check specific gravity using hydrometer
- Are terminals tight and clean? If not remove, clean, lubricate with silicone grease and replace tightly. Check cable to terminal connections
- Clean batteries, especially on top to avoid leakage between terminals and also battery tray
- Are batteries secure?

Alternator
- Are all terminals tight and clean? Use Waxoyl or similar carefully to protect terminals
- Spray carefully with WD40 or similar to remove moisture
- Check drive belt tension and tighten if necessary
- Check for bearing wear by running engine slowly and listen for familiar whirring/whine sound

Switches, terminals and plugs
- Operate all switches, both on panel and individual appliances. If sticky, loose or ineffective either dismantle to check, clean (WD40 useful again) or replace
- Check all terminals for tightness and coat with spray wax (Waxoyl again)
- Check all plugs, especially deck plugs for corrosion. Clean and coat with silicone grease. Self amalgamating tape offers extra protection for deck plugs

Wiring
- Check accessible cable covers for damage and chafe, especially where they pass through bulkheads, etc
- Minor chafing can be dealt with by parcelling cables with tape
- Cables that are seriously chafed (ie through or nearly through the outer sheath) should be replaced

A manually-operated four-way battery switch directs the current from the alternator to the battery you want to use. The switch controls both where the current will be drawn from, and where the charging current is to be directed. Although the wiring is simple, the most important feature is that such a swith must 'make' before it 'breaks'

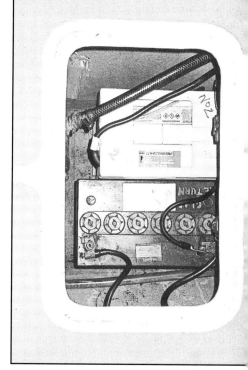

Batteries of different types and capacities should never be charged together nor be left in service when connected in parallel. A low stowage position such as this is fine so long as they can't be reached by bilge water

being deeply discharged.

Installation. Batteries are heavy items specifically referred to in the Fastnet Report and the ORC Special Regulations. They must be securely restrained not just for the safety of the crew but also for their own good. A strong box with a lid which can be screwed, strapped or locked down should be provided. The box must be drained and ventilated to remove water and gases. Extra care should be taken to mount sealed batteries in a strong and secure box as they have been known to explode.

The batteries should be positioned close to the engine, to reduce cable runs to a minimum, but outside the engine compartment itself where heat and vibration can cause damage. Huge losses can take place here if cable runs are too long. It is not just a question of longer runs slowing the rate of charge to the battery, but of actually preventing the battery from reaching full charge.

In order to monitor the performance of your electrical system it is a good idea to fit an ammeter and voltmeter so you can check on rate of charge and battery state. Even so, it is still hard to beat the old fashioned hydrometer, though with sealed batteries this is not possible to use.

Supply

If you thought you had problems getting electricity into the battery in the first place, it is nothing to the difficulty of getting a decent proportion of it from the battery to the relevant appliance.

The wiring on boats is notorious for the toll it takes on current. This is partly due to the inherent drawback of low voltages and therefore high currents, to the length of the wiring needed to get current, not just from one end of the yacht to the other, but to the top of a mast perhaps 20 per cent longer than the boat. The cable run from switch panel to truck and back again on a 35ft boat may be as long as 120ft.

From the battery the first stop should be a properly designed switchboard fitted with a battery isolation switch. Each individual circuit should be fused, or better still fitted with a circuit breaker of the correct rating. This will make fault finding easier and circuit breakers are sometimes fitted with neon or LED indicators to show when they are in operation, sometimes coupled to an outline of the boat to show what navigation lights are in use. Perhaps this trend will encourage the correct use of lights at night.

The distribution system should be insulated return – ie the negative is not grounded to the engine but instead completes a circuit back to the battery – using wiring of the correct cross sectional area. Voltage drops of no more than half a volt should be tolerated. Correct cable sizes and how to calculate the right size for an appliance are illustrated in **Fig 4**. Because wires on a yacht can move and so lead to the possibility of fatigue, a multi-strand cable (flex) as opposed to single

Adding electrical equipment

If electrical equipment that uses a significant amount of power (either individually or collectively if several items are being fitted) is added to an existing system, then it is possible that the supply wiring to the switchboard, battery and even the alternator may need uprating. With a healthy battery(ies) and a standard alternator this should not be necessary if the total new requirement is less than two or three amps. However for items such as heating, radar or a fridge it is recommended that professional advice is sought.

Gear with low power drain (Decca, Autohelm, etc), unless complex wiring is needed, is usually simple to fit, though care should be taken to ensure good connections at the switch panel

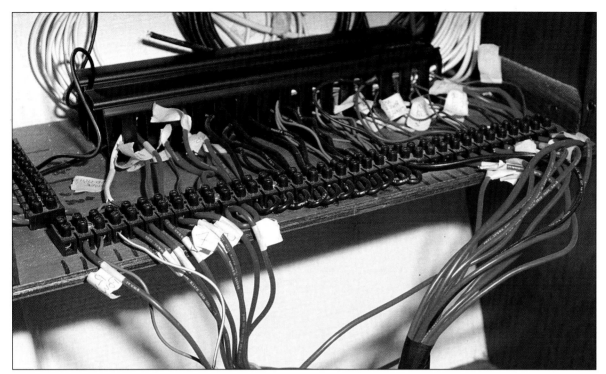

There will come the day when you will praise the builder who clearly identifies wiring and curse the one who does not. The same goes for the electrician who considerately leaves enough cable slack so that the switch panel can be removed easily

Electrical troubleshooting

Electrical faults are notoriously difficult to trace and your most valuable companion will be a multimeter, ideally one with digital read-out that also acts as an ammeter. A straightforward and logical approach is the only answer, checking circuits and appliances one by one until the faulty one is isolated

strand domestic wiring should be used.

The biggest voltage drops invariable occur when connections are badly made. A few strands broken off can severely curtail the ability of a wire to carry electricity while corrosion and damp also cause losses. Crimped terminals provide the best connections to equipment but the crimps must be firm and made with a proper crimping tool as opposed to a pair of pliers. These connections can be tested by simply pulling the wire and looking for movement.

Where strip connectors are used they should be securely mounted. Wire ends should be soldered to create tidy efficient contacts within the connector but soldered joints between wires themselves are not advisable in a marine environment because bimetallic action can lead to corrosion. Various forms of proprietary connectors are available including the 3M links (from car accessory shops) which can be made without removing the insulation cover. 'Chocolate box' type of connectors can be perfectly satisfactory as long as it is remembered that they are likely to use mild steel and suitably protected with wax or completely sealed with silicone sealant.

Cable runs should ideally be inside plastic pipe conduiting. A piece of line left inside the conduit will make the addition of extra wiring later much easier. Loose wires running together should be bundled at regular intervals and secured to the fabric of the boat. Avoid running wires down both sides of the boat as this creates closed loops leading to radio interference.

Where cables have to pass through the deck, waterproof plugs should be used for mast bound cables and deck glands for other wiring. The supply of electricity to navigation lights and other remote exposed equipment is notoriously unreliable. The life expectancy of these circuits can be greatly improved if proper connectors are used, connections are carefully made and protected, cables run through conduiting to prevent chafe and the correct cable size used. For protection, deck plug and socket terminals should be coated with silicone grease whilst most other terminals are best protected by a wax film as opposed to an oily spray. Finnegan's Waxoyl, available from car accessory shops in aerosol form at reasonable cost, is excellent for this and leaves a thin coating.

Conclusions

Whilst it may be argued that electrical systems are not wholly essential to the ultimate seaworthiness of the modern yacht, without doubt they play an important part in the safe running of a yacht at sea and are a great deal more than just convenience. Properly installed, protected and maintained there is no reason why such systems should not prove reliable and continue to operate in difficult conditions.

Fig 4 Cable rating table

The correct cable can be calculated by the taking the maximum allowable voltage drop (0.5v for a 12v system) dividing it by the current required by the appliance (say 10amps). This gives an allowable resistance of 0.05 ohms. In order to use the table multiply this by 1,000 to give the cable resistance rating per 1,000m and then divide it by the total length of the cable to be used (there and back), say 10m. This gives a figure of 5 ohms. From the right hand column the table shows that a cable of cross sectional area 4mm^2 is required. This is much larger than that suggested by simply looking at the central column for the cable rated nearest to 10amps (1.5mm^2).

Cross section area mm^2	Rubber and PVC covered		
	Single core amps	Two core amps	ohms
1	8	7	18.84
1.5	12	10	12.57
2.5	17	14	7.54
4	22	19	4.71
6	29	25	3.14
10	40	34	1.82
16	54	46	1.152
25	71	60	0.762
35	87	74	0.537
50	106	89	0.381

Chapter 10

Equipping for emergencies

However well found and equipped a yacht, there will be occasions when she will encounter problems that may develop into full-blown emergencies. The philosophy of equipping for safety at sea should be one of preparation for problems first and only then considering the escape route, the remedial emergency gear. This chapter examines priorities for safety

CLOSE YOUR EYES for a second and ask yourself a simple question. What colour is safety? You'll most probably reply yellow or orange. It's not. Safety, if it has any colour, is grey and lives between the ears of a yacht's crew members. *Emergency* or survival gear might be yellow or orange but the implication is that, if you're that far down the line in an emergency, at some point your safety defences have been breached and the situation has passed out of your control.

It is the terminology used in association with emergency equipment that is partly at fault. We have liferafts, lifejackets, lifeharnesses and fire extinguishers. Better descriptions would be survival raft, emergency flotation, safety harness and fire appliance because they reflect their function more accurately and what a crew's expectation might be of them.

True safety begins at the far more fundamental levels of anticipation and prevention and it is the understanding of these that will lead naturally to equipment priorities and the proper appropriation of funds to safety and emergency equipment. Whilst it would be irresponsible to suggest that yachts put to sea without basic emergency and survival equipment, if these long stops are removed from the safety equipment list when planning for emergency situations, it will be easier to recognise and clarify the priorities.

The first step in prevention is recognition of the problems, which are easily remembered as the six Fs:
Flooding
Foundering
Fire
Fog
Falling (people and masts)
First aid
If you ask yourself next 'If it happened to me, how would I cope?' you are well on the way to prevention.

Flooding
There can be two types of hole in a yacht's structure; those that are meant to be there and those that are not. The intentional holes are the openings for windows, hatches, water inlets and outlets and through-hull instruments. The unintentional are those caused by accidental damage to hull or deck.

Intentional holes. However strong and well fitted, you should always consider the possibility that deck openings or through-hull fittings may fail. You may double clip and service seacocks religiously and check locker hinge fastenings to reduce the chances but these will never be totally eliminated so emergency measures must be planned and the right tools and materials carried on board for repairs.

All through-hull plumbing ought to be double clipped both at the seacock and at the other end, taking special care with engine inlets as they lead to a cooling water piping with several connections, any one of which might fail. If you've a hot water system don't forget the calorifier either.

Tools and materials for coping with flooding via hull openings are simple: a few assorted corks and rags, tapered wooden plugs which are available from any chandler, moisture-cured sealant (LifeCalk is excellent), various Jubilee clips, a hammer, screwdriver, pliers and adjustable spanner.

Failure of any of the windows or deck openings, although above the waterline, are potentially more dangerous purely on account of their size. In the 1979 Fastnet, 42 per cent of the competitors who returned questionnaires considered that companionways were a significant water entry point (one yacht sank after losing her hatchboards and being flooded) and 20 per cent said the same of cockpit lockers, yet none reported the failure of through-hull fittings. The loss of a cockpit locker lid of the size seen on many modern production yachts in heavy weather could be catastrophic unless the means to repair it are carried – on board.

The simplest emergency repair kit would contain pre-cut sheets of ply to cover the largest openings, lashings of sealant or underwater epoxy putty, sharp drill bits, a good supply of self-tapping screws and a panel saw. For windows, a piece of wood cut

New since 1979

For many years, the search and rescue agencies have been working internationally to establish the GMDSS – the Global Maritime Distress and Safety System

GMDSS identifies a number of key elements in any search and rescue incident, including raising the alarm, establishing a search datum and localisation by a rescue vehicle.

The first two elements are dealt with by the COSPAS/SARSAT satellite system, detecting and fixing the position of a signal from an emergency position indicating radio beacon (EPIRB). The satellites can receive on two frequencies, 121.5MHz and 406MHz.

121.5 was originally (and still is) an aviation distress frequency. EPIRBs operating on this frequency have been available for many years and relatively inexpensive ones, some costing under £100, are now available.

There are two drawbacks with 121.5 beacons. To work effectively through a satellite, the beacon must be within the satellite footprint, the area in which the satellite is above the horizon both to the beacon and to an earth station. On this frequency, the satellite simply receives the distress transmission from the EPIRB and rebroadcasts it immediately. Fortunately, there is an extensive footprint around the British Isles but there are a number of ocean areas of the world where 121.5 beacons cannot work through satellites.

The second drawback with 121.5 beacons is their very poor record for false alarms, well illustrated by recent figures from the Department of Transport: during the period January to June 1988 there were 430 COSPAS/SARSAT alerts on 121.5 Of these: 271 were identified as spurious transmissions due to atmospherics; 159 became 'incidents', ie transmissions were picked up by subsequent passes of the satellite. Of these: 102 had no final outcome, despite monitoring for 24 hours. 50 were subsequently traced to beacons inadvertently switched on. 7 were found to be true emergencies.

The 406MHz beacon is much more sophisticated. It transmits coded information, which can include nationality, type of vessel and beacon serial number and there is a facility to indicate the nature of the distress. Signals received by the satellite are, if necessary, stored until they can be transmitted to an earth station, so there is world-wide coverage.

The satellite position fixing facility on 406 is also superior, to within 3.1 miles, compared with 12 miles on 121.5. This may seem a fairly trivial difference, until one realises that the resulting search areas are of 20 square miles on 406, compared with 452 square miles on 121.5.

No single system provides everything the user could possibly want and the drawback with 406MHz beacons is their price. The least expensive one currently available costs over £800.

It's a dangerous fallacy to think that buying emergency equipment will make you any safer

Two part epoxy putty both sticks and cures underwater

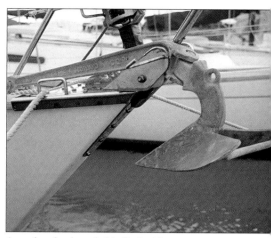

A good anchor is probably the best insurance you can buy. Don't skimp by choosing cheap variants, make sure you buy an anchor with a proven record

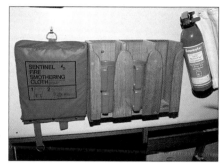

Extinguisher and fire blanket in the galley are to hand to cope with small cooker flare-ups but it is highly unlikely that you will be able to tackle established fires with on board equipment. Your priority should be to get the crew off the boat

We all worry about hull damage – in bad weather the loss of a cockpit locker lid of this size could be equally catastrophic

Everyday items such as odd pieces of timber and sleeping bags might provide the first line of defence if holed

Gas is the most dangerous fuel you have on board. Check and double check all connections and piping for leaks. A Gaslo meter (just visible on the top bottle) will give an early indication of pressure drop in the piping

a bit larger than the largest can be prepared with two strongbacks pre-drilled and with bolts kept in position.

Sailbags, with sails in them stuffed rather than folded, make good temporary covers for large openings, perhaps just whilst a piece of ply is being cut to fit.

Unintentional holes. It is unlikely that the crew of the average yacht will have the wherewithal or the skills necessary to repair a large hole below the waterline at sea.

However, there is quite a lot that can be done to prepare for the possibility of a smaller hole. You are looking for two levels of remedy: first aid to slow the water flow and the get you home repair. For the former the materials and tools you should consider are: an axe for clearing away internal furniture in the area of the breach; sleeping bags and clothes to stuff into it – or possibly one of the purpose-made Subrella patches; shores to hold patches/stuffing in position. It is unrealistic to expect every yacht to carry long lengths of timber for shores, but it should not be necessary if you've got oars, boathook and booming out pole on board.

For a longer term repair, you might consider the following. A small bag of quick drying cement and a large Tupperware box, clearly marked for emergency use, which might contain underwater epoxy putty, a selection of self-tapping screws, a sharp drill bit wrapped in an oily rag, a tube of polysulphide sealant and a selection of rags.

For a patch, unless you carry odd bits of ply, you've the floor and hatch boards and locker lids. Once you've stemmed the initial flow, you can prepare a patch with a dozen or so small knobs of epoxy putty round the edges, holes drilled in the corners and a shore ready to hold it in place when you remove the first aid covering.

You might consider also rigging the storm jib outside the hull as a collision mat, lashed around the boat like a bandage but the shape of the boat and the time needed to set it up may make this impractical.

Coping with flooding. Unfortunately very few production boats have bilge pumps of adequate capacity to cope with serious flooding. And along with water swilling round the boat will be matches, bits of paper and all sorts of debris ready to block the pump. It is well worth considering large capacity bilge pumps as well as an engine driven pump.

The old saying is that there's no bilge pump as effective as a frightened man with a bucket. Peter Haward suggests a variation on the theme by commenting that nothing's more effective than a frightened man with a dud VHF and a bucket. The bucket should be the heavy duty black, agricultural type with metal handles.

In any emergency, a VHF radio may prove your lifeline – far more effective than flares in range and efficacy. And yet if a boat is flooded it will be one of the first casualties when the batteries are immersed. An emergency VHF power supply, perhaps a small battery mounted under the deckhead, is worth consideration just for this reason, as is a hand-held VHF.

Man overboard – prevention should be the keyword and safety line eyes fitted in the cockpit with jackstays on deck

If you lose the mast, have you the wherewithal for a jury rig? A length of rigging wire and some bulldog grips are essential emergency spares

The panic bag

The contents of the panic bag will depend how long rescue might take, so the blue water yachtsman will have very different requirements to the Channel or North Sea sailor. The list below is intended for the crew of a yacht who can realistically expect rescue in less than 48 hours.

Container: a watertight container (possibly two) such as those made by PCM Ltd (Tel: 09285 75051). This, when loaded, should float and not be so heavy that it is not easily lifted. It should also have a strong point and long lanyard attached with snap hooks *at each end*.

Warm, dry clothes: including gloves, thick socks and woolly hat

Space blanket(s): as sold in camping/climbing shops

Water: enough for two pints/day/crew member carried in a separate container or mineral water bottles

EPIRB/waterproof hand-held VHF: or one kept in a waterproof pouch

Flares: Parachute and hand-held reds, smokes. Enough parachute flares should be carried to allow them to be fired in pairs, a couple of minutes apart

Torch(es), whistle and heliograph

Knife: (Swiss Army type) and assorted line

Seasickness pills and basic first aid kit: include specific personal medicines and Vaseline for salt water sores

High energy snacks: Mars Bars, Kendal Mint Cake, glucose sweets

Comforts to keep up morale: pack of cards, books, notepad and pencil, sponge

Flares are fine...provided they are in date

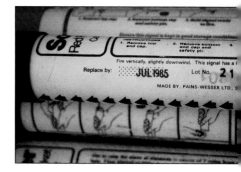

Foundering

Land is the yachtsman's biggest enemy. We might all fear the semi-submerged container on a dark night but overwhelmingly it is unintentional grounding or hitting rocks that causes most total losses.

It is for this reason that a good anchor is the best insurance you can buy for your boat. If you are drifting on to a lee shore it is the anchor that offers the last line of defence. If aground and pounding, an anchor laid off may prove your salvation and it is a false economy to skimp on weight or cable size, as it is to buy a cheap anchor as opposed to a proven type.

But if you are holed and unable to stem the flow of water, and the only avenue left is to abandon ship, you should have thought through this scenario carefully beforehand. Do you and your crew actually know how to launch the liferaft? Do you know what is inside it? Have you thought about what else you might take with you when you abandon ship? Any yacht going offshore should have a panic bag prepared in recognition of the primary problems. Survival comes first with rescue a close second.

In northern European waters, hypothermia is the main threat, even in summer. Most people can last for several days without food, but not without water, so this should be the next priority.

Turning to rescue, how about flares, an EPIRB or portable VHF? In the accompanying table we list what your panic bag might typically contain.

Recent years have seen the development of cruising yachts with positive flotation achieved by a foam filled double skin construction, notably the Etap and Sadler ranges. Larger yachts can be similarly protected by the sub-division of the hull into separate watertight compartments, though for boats less than 40ft this is usually impractical. One possible solution is the use of inflatable flotation bags, normally stowed below bunks such as the Unsink system made by Anglo Dutch Engineering (Tel: 01 686 9717).

Even if positive flotation is achieved, a new problem presents itself: that of many tons of water surging to and fro in the hull. As water runs forwards, the pressure it develops could be enough to blow out hatches or even lift the deck so another item to consider is some kind of emergency arrangement to fit baffles, perhaps sheets of ply, to contain the water and possibly isolate the damaged compartment.

Fire

Boats are highly flammable. We carry dangerous fuels on board, some explosive, and unless specially treated the very fabric of a yacht, glassfibre or wood, burns well. Once a fire has taken hold, there is very little a crew armed with a normal complement of extinguishers can do to put it out. The keyword here has to be prevention.

Prevention means constant checking and vigilance, replacing even slightly suspect components in a fuel or gas system, strict rules for handling flammable liquids and clearly understood routines for dealing with leaks.

In terms of equipment, the least you might consider is a remote gas tap by the galley, a Gaslo indicator on the bottle and perhaps a gas detector, though these should be tested regularly with a gas lighter. Electrical fires are the most common on board which means checking wiring where it might chafe and replacing old and brittle circuits. Every boat, whatever her size, should have a *minimum* of one extinguisher per cabin of a minimum size of 2kg (Halon and dry powder) and also fit a small unit by the galley in addition to a fire blanket to nip cooker fires in the bud. Purpose made fire blankets are only intended for small fires. For larger fires it can be effective to use a domestic blanket soaked in water to smother the flames.

Fog

The only guard against fog is not to sail when it is forecast. But you should be

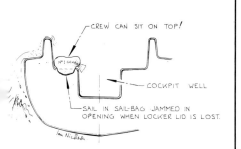

You know how difficult it can be to stuff sailbags into cockpit lockers. If a locker lid is lost you can use that to your advantage by stuffing the hole with a loosely bagged sail

prepared against the eventuality of being caught out in fog and understand the dangers fully.

In pre-Decca days, the uncertainty of navigation kept crews alert to the real danger, that of collision. Today, such is the trust in electronic navigators that the temptation is to think that the danger of fog has diminished. If anything, the opposite is true as the risk of 'Decca assisted collision', of two yachts using the same waypoints but in opposite directions, has increased.

Fortunately, most of us are provided with reliable anti-collision aids as standard, our sight and our hearing. For a few pounds an aerosol foghorn can be purchased, though there is a tendency for the reeds in them to freeze (warm over the stove between toots). But the effectiveness of these when confronted with a ship's watch officer in an enclosed wheelhouse is probably more as a morale booster for your crew.

Being seen has to be priority number one which means using navlights in fog and hoisting an effective radar reflector at the first hint of deteriorating visibility. By effective we mean an octahedral, in good condition and with a minimum size of 18in across the diagonal, properly hoisted in the catchwater position (ie with a cavity, *not* the point, up) or a reflector of the Firdell or Gillie Firth type with proven performance data. Don't be tempted by any patented devices unless their performance claims are backed by independent test results.

Radar detectors can warn of vessels using radar in the vicinity though not all indicate direction. However, the best anti-collision aid there is, provided you can afford *and know how to use it*, is radar. Earlier objections to small boat radar on account of scanner size, weight and the power consumption have largely been overcome with the new generation of compact sets from Furuno, Raytheon, Vigil and Apelco.

Falling (people)

'If you fall overboard in open water consider yourself dead.' This might sound extreme, especially in view of the advances made in man overboard location and recovery systems in recent years, but the sad fact is that in a significant proportion of man overboard accidents the casualty is not recovered.

Although man overboard systems will increase the chances of recovery, equipment priorities should be first and foremost on prevention. This means easily used harnesses for every crew member, harness eyes in the cockpit and near open areas of deck, and jackstays accessible from the cockpit that allow crew to move the whole length of the deck without unclipping. It also means effective non-slip on all deck areas, including the coachroof, and non-skid soles on boots and shoes. And none of these is any good at all unless there are clearly understood routines and disciplines for using harnesses.

Personal preferences usually dictate the type of harness used. Each has its pros and cons. Some prefer them built into foul weather jackets so they are always worn when the going gets rough; others prefer separate harnesses. Above all, they must be easy to adjust and put on (especially in the dark); otherwise the chances are that they will not be worn in marginal conditions.

Unfortunately, few harnesses carry the BSI Kitemark – not because they don't comply but because the cost of testing adds between 10 per cent and 15 per cent to the price which, in a small and competitive market, is a major disadvantage. Many, however, are made to BSI specifications and the features you should look for are outlined in the accompanying table.

If someone does go overboard his first concern – and yours – is that he should stay afloat. In reality very few crews wear lifejackets at all times. They are seen as survival equipment to be worn after an emergency has occurred or perhaps in fog. This is nothing to do with macho risk taking; it is the sheer inconvenience of wearing them. Automatic inflating jackets are less bulky and combined lifejacket/harnesses such as those made by Crewsaver are another answer. But the chances are that, when that crew member goes over the side, he or she will not be wearing a lifejacket.

This means that the main line of defence is the lifebuoy. Under the heading Basic Requirements, the ORC special regulations state clearly that *'all required equipment shall: function properly, be readily accessible, be of a type, size and capacity suitable for the intended use of the yacht'*. The same regulations demand one lifebuoy with drogue and light attached and another with a whistle, dye marker, drogue, light and danbuoy.

That's an awful lot of string to get tangled on a dark night, as one Fastnet Race crew discovered in this year's race. When a crew member went overboard they could not launch a lifebuoy because the lines tangled. It took quick thinking and a sharp knife to cut the line and throw over the strobe light. He was finally recovered nearly an hour later.

There is a strong argument for simplicity, for one of the lifebuoys to be unencumbered except for a drogue, for individual crew to carry whistles and for the danbuoy and light to be launched separately. It might leave your wake littered with several separate items but that's infinitely preferable to the whole lot tied together with a ball of knitting and still in the cockpit.

Location is the second problem and every serious cruising yacht should carry at least two lifebuoy lights, a danbuoy with light or retro-reflective strips, a powerful hand-held spotlight and dye markers.

The final and by no means the least problem is recovery. On a fully crewed yacht sheer brute force might be adequate but on a lightly crewed cruising boat some means has to be arranged to get a casualty back on board. It might be a purpose-made tackle, a spare halyard led to a winch, a ladder or the inflatable but it must be possible for the weakest crew member to recover the heaviest.

There are a number of promising developments which may have real value for location and recovery. For some years devices like the Lifeloop recovery/lifting pole have been on the market as have several clever ladder and scrambling net recovery aids. More recently we've seen the Jon Buoy

Safety harness

This list is a précis of ORC minimum standards which in turn are based on BS 4224-1975 and AS 2227-1978. For the full ORC Special Regulations send £2.20 to Offshore Racing Council, 19 St James's Place, London SW1A 1NN.

- The design must allow the safety line to be located at armpit level. It must be adjustable but any adjustment buckles must not slip under load
- Chest strap should be a minimum of 38mm wide, brace straps (not usually fitted to integral jacket/harnesses) not less than 19mm
- The safety line should not be longer than 2m with a hook at each end. Any intermediate hook should meet the same strength requirement as those at the ends (minimum 1,500kg breaking strain, SWL 700kg)
- Breaking strain of the safety line should be not less than 2,080kg and of the harness webbing not less than 1,000kg per 25mm width
- Splices in the safety line should have at least four full tucks and two tapered tucks, and be whipped or otherwise protected

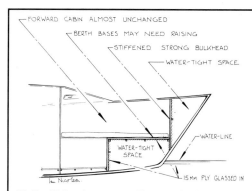

Hull collision damage usually occurs in the bow sections. Even quite small boats can be easily protected by creating watertight compartments and collision bulkheads without major inconvenience

which releases a mini raft with integral lifting sling. The makers of the Jon Buoy, Trans Aqua Technology, are currently carrying out further trials with the idea of linking an automatically-released buoy or line to a yacht's Decca and autopilot, recording an emergency position on the one and locking the other hard over to stop the boat.

The emergency position or man overboard button on Decca itself is, of course, useful for location whilst Navico have recently launched a remote keypad for one of their autopilots which gives course and position back to the point where the man overboard function was activated.

Falling (masts)

When a rig goes over the side, there will be very little that can be done to save any part of it. It is highly unlikely that a cruising crew will have the strength to recover a waterlogged mast of more than 30ft in length and at best they may be able to lash it alongside in the hope of salvaging parts of it later.

So the first priority should be to cut it free before it batters a hole in the hull. For this, you'll ideally need large 'pincer' bolt croppers of the type made by Felco. Additionally, a large hacksaw with plenty of spare blades should be carried plus a couple of pairs of heavy duty pliers for removing reluctant cotter pins. On the subject of cotter pins, retaining split pins should never be bent right back on themselves as it makes them very difficult to remove in a hurry. They need only to be splayed out 10 degrees or 20 degrees to hold them in place.

Heavy duty gloves will prevent injury from sharp metal edges and broken wires, and a mole wrench is often faster for cutting through a mast than a hacksaw – it is used to 'nibble and bend' the aluminium until it fatigues and breaks.

Once you've cut away the damaged rig, how are you going to get home? Spare rigging wire, hard eyes and bulldog grips will prove valuable when setting up a jury rig as will an emergency VHF aerial to let the world know that you are getting home under your own steam and, no thank you, you don't need any help.

And if you decide to use the engine, don't forget to carry a mask and snorkel which will be needed when you go swimming to free the propeller that has fouled when you forgot to check that all lines were inboard before starting.

First Aid

There are a number of excellent purpose made yachtsman's first aid kits on the market or, alternatively, if you explain the requirements your GP should be able to advise on suitable contents for your own kit. As important as the kit is the knowledge of how to use it with a first aid manual. The Royal Yachting Association produces an excellent one (publication G6, price £2 from the RYA, RYA House, Romsey Road, Eastleigh, Hampshire SO5 4YA).

The commonest injuries suffered on board are minor cuts, bruises, abrasions and burns, and the shortcoming of many first aid kits is in the under-provision of common or garden bandages and creams to treat these.

What flares for offshore?

ORC Category 2 recommends the following minimum for flares. These must be in date – that is within three years of the stamped date of manufacture and stowed in a waterproof container

- 4 red parachute flares
- 4 red hand flares
- 4 white hand flares
- 2 orange day smoke signals

Additionally, you might carry a mini flare pack (useful for the pocket of crew on watch) and your recently time-expired flares, but never fire out-of-date parachute flares that look in any way damaged.

The final resort. But if your boat is still afloat does an 8ft inflatable offer a better chance of survival?

There is always the risk of a major injury but medical problems are far more likely to stem from deterioration of improperly or inadequately treated minor injuries.

We've all heard tales of the singlehanded doctor who removed his own appendix at sea. The reality is that if you have a major medical emergency on board, you will be unable to cope, even if you carry the equipment to stock an operating theatre, unless specially trained, so you are going to have to call for help on the VHF. Your kit should reflect this fact and be aimed at primary first aid until help arrives. Plasters, bandages, dressings, antiseptic and burn creams and aspirin there should be in plenty along with the Stugeron and hangover cure.

Finally...

Quite intentionally the liferaft has been left until last. If you've got this far in your anticipation and preparation for emergencies then the chances of needing it are much reduced. A detailed analysis of the modern liferaft appeared earlier on Chapter 8 so we will add just a few more thoughts.

Size of raft. If you have the misfortune to take to the liferaft you'll find it very cramped with a full complement. But don't be tempted to go for one larger than needed as human body weight forms a vital part of its ballast.

Do you know how to use it? Take the opportunity when yours is next serviced to go along and inflate it yourself and examine its contents.

Abandoning ship. Unless forced to do so by other circumstances, don't board the liferaft until you are absolutely certain that the yacht is sinking. A 30ft waterlogged yacht is a far better liferaft than an 8ft inflatable.

Alternative rafts. Several companies, notably the makers of the Tinker range of inflatables, offer survival options including CO_2 inflation, canopy and drogue. The Tinkers have been tested extensively and been used for real. However, because there is not the same servicing requirement as for a liferaft this puts the onus on you to ensure that the inflation system still works and that the whole is still in serviceable order.

Chapter 11

The Yachting Monthly Offshore 35

*Throughout this book we have examined every aspect of detail and design that makes for seaworthiness and a yacht fit for offshore sailing. To conclude we have taken all of these and incorporated them into one yacht, the **YM Offshore 35**. Illustrations are by John Moxham and Arthur Saluz*

THIS IS NO idle design exercise or concept yacht. Nor do we claim to have broken new ground with revolutionary design, construction or technology. What we have done is to take a set of design parameters that will create a boat with desirable stability, handling and performance characteristics and then build into it all the very best and proven features and equipment.

The rationale for the design is that the seaworthiness of a yacht is dependent on the total sum of her parts. A hull shape alone does not generate seaworthiness, nor does the efficient working of the yacht. All seagoing yachts are made up of similar basic components and gear, but it is the relationship between these that will make one yacht a dream and another a dog.

To define these parts we have relied on our own extensive experience. We have culled workable and practical ideas from 300 or so different yachts. And last, but by no means least, the initial outline concept was presented to our Advisory Panel, whose enormous combined experience as seamen, designers, in yacht research and in surveying is unrivalled, for discussion in a forum. This is the result.

YM OFFSHORE 35
Basic design parameters

Ballast ratio	42% to 45%
Length/displacement ratio	225 to 250
Length/beam ratio	2.6 to 2.8
Beam/depth ratio	5 to 5.5

These led to the following dimensions. A length of 35ft (like all good ideas, it grew a bit) was selected on the basis that any smaller yacht would be unlikely to conform with Chuck Paine's useful 20/20 rule (full sail carried in 20 knots of apparent wind would, to windward, result in no more than 20 degrees of heel). However, it must be said that most of the features and details discussed here are equally applicable to smaller or larger yachts.

LOA	35ft 6in
LWL	29ft
Beam	10ft 6in
Draught	5ft 10in
Displacement	13,000 lb
Ballast	5,600 lb

The Wolfson Unit ran the figures through their computer which predicted a **minimum** range of positive stability of 140 degrees, which implies that, should the boat be knocked down to 180 degrees, the chances of her remaining inverted for more than a few seconds would be minimal.

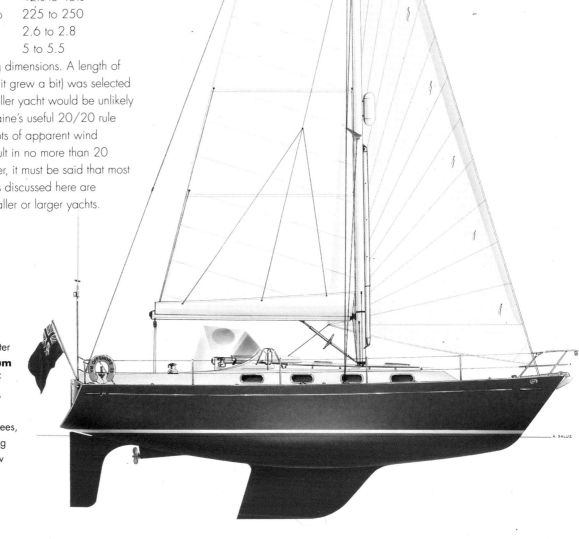

YM OFFSHORE 35

- The headsail roller drum is below deck level at the forward end of the anchor well for maximum luff length and to keep the deck clear for anchor handling
- All lines and reefing pennants lead aft, running below hinged garages to keep the deck clear
- Halyard and reefing line tails stow in open-topped coaming bins in the cockpit
- The dorade ventilator hoops serve to prevent lines catching, to protect the cowls and to act as grab bars
- The pulpit 'goalpost' keeps the antennae out of the way and allows for better reception. Autohelm vane, stern light and cockpit light are also mounted here
- The cockpit drains via a wide transom slot. The ladder folds down from the scoop and the dan buoy stows in an internal tube

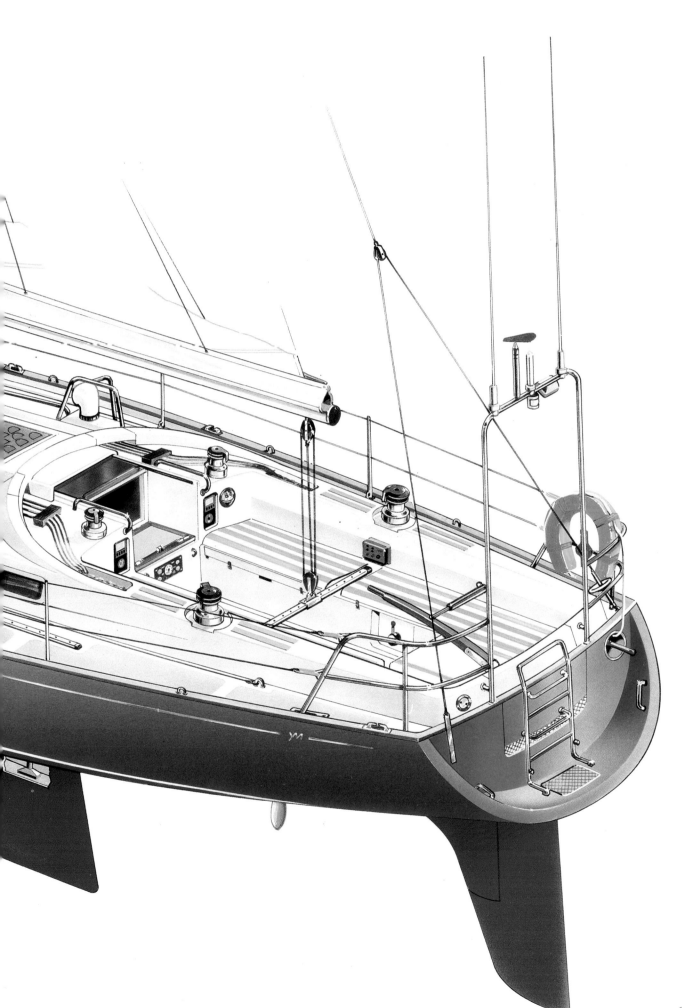

Hull design and structure

Compared to many of today's production yachts, the YM Offshore 35 has not been designed to create maximum internal volume for in-harbour living. Although she has a comfortable accommodation, the prime requirement was to move the emphasis back towards seagoing ability. To this end she is long on the waterline (29ft) for speed potential, and is relatively narrow (10ft 6in) to be easily driven and pleasant to steer when heeled, although the beam is carried well aft giving quite a broad transom. She has deep, veed forward sections, a fairly deep canoe body and a clean run aft with slightly veed sections.

She should also be stiff because her lead keel is slightly bulbed at its base to lower the centre of gravity. The ballast is fitted to a deep keel stub which creates internal space for a respectable bilge sump and also to allow for the future provision of a holding tank. The toe of the keel is well rounded to minimise potential damage on grounding and for the same reason the base of the keel has a slight lift from aft to for'ard. The propeller shaft exits via a shaft log as opposed to a P-bracket, whilst the rudder is partially balanced and mounted on a substantial skeg both for mechanical strength of the lower bearing and also to aid directional stability.

Both the freeboard and the height of the coachroof have been kept to modest proportions but still allow over 6ft of headroom in the main parts of the interior. As with many modern designs, the transom is inset to create a scoop for boarding from a dinghy with a fold-down bathing ladder. Handholds are fitted here for holding on from a dinghy. As the boat would be for series production, construction would be conventional GRP, although the design would lend itself to one-off wood epoxy building.

Watertight integrity

It became apparent during the course of this research, and in particular when analysing the 1979 Fastnet Report in detail, that lack of watertight integrity quickly started a chain of possibly disabling problems.

To this end, the YM Offshore 35 has no openings in the topsides and all her coachroof windows are kept to a small area (within the ORC regulations). None of them opens. Ventilation, both at sea and in port, comes from centreline opening hatches, dorade vents aft and a Tannoy forward, with a large aftercabin hatch opening into the cockpit well.

The perennial companionway problem of what to do with washboards and how to use them in heavy weather is neatly solved with a single hinged washboard (see **Fig 1**), comfortably larger than the aperture and permanently attached to the yacht.

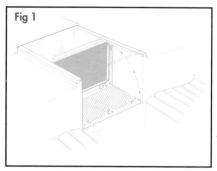

The single washboard folds flat onto the bridge-deck solving the problems of stowage and security. Note that the hatch, when fully closed, comes right to the after edge of the coachroof to allow ventilation on rainy days whilst keeping the interior dry. The seats are contoured, park bench fashion, for comfort

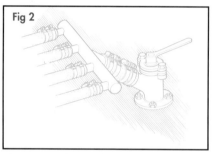

There are only two through-hull openings, an inlet by the galley and outlet in the heads. All piping to and from services lead to manifolds on the seacock, with one-way valves on the outlets. All piping is double-clipped

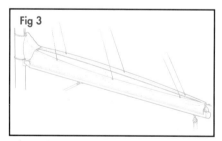

Stackpack mainsail system. This consists of a fabric pouch, secured in grooves either side of the sail foot and supported by a batten along the top edge and lazyjacks. Once the sail is dropped into the pouch it is zipped shut and it also holds the sail in place when reefed. A small bonnet at the gooseneck is fitted for in-harbour stowage

The cockpit itself is drained straight out through a wide transom slot. The well is slightly wider at the stern to assist drainage when heeled over. This system is simple, very efficient and avoids extra seacocks/plumbing. Cockpit locker lids have been kept small.

Another favourite water ingress point has been eliminated by having the anchor chain in a self-draining well forward, sealed from the remainder of the accommodation by a substantial bulkhead that also acts as a collision bulkhead.

There are only two through-hull fittings, one inlet by the galley and one outlet in the heads, each fitted with a manifold (see **Fig 2**) to distribute the plumbing to different services. These are all double clipped and have one-way valves fitted.

The rig and sails

We opted for a deck-stepped, fractional rig for a number of reasons. The YM Offshore 35 was to be suitable for family sailing and in this respect we liked the smaller headsail. We were also keen, as far as was possible, to have an all-round headsail which operated efficiently throughout a wide range of windspeeds without the need for a suit of headsails (despite the roller).

The modern mainsail, especially with full battens and stack-system lazyjack/cover, is very easy to handle throughout all wind ranges. It therefore 'deserved' a greater proportion of the total sail area compared to the headsail which, whilst set on a roller (within the anchor locker), is not as adaptable in varying conditions. However, by having the tack aft of the stem, there is a facility to set a light ghoster loose-luffed forward of the forestay (rolled away through each tack). The existing headsail is shaped and proportioned to work well in light airs right through to being well reefed in anticipated 35-38 knots apparent windspeed. The angle of the forestay, high clew, radial-cut and foam luff will make this jib an excellent reefing sail without the need for much sheet track adjustment (which is done from the cockpit with car-tackles). Above 35-38 knots apparent, the yacht would either carry on under deeply reefed mainsail only (another advantage of the fractional rig) or set a hanked-on storm jib on a separate removable stay immediately behind the existing forestay.

The mainsail uses the 'stack-pack' system (see **Fig 3**) pioneered by Doyle Sailmakers, since emulated by a variety of sailmakers and adopted by builders such as Dehler and Etap. A fixed rod kicker (rather like a very large bottlescrew) with adjuster wheel is employed to cut out two control lines – the kicker and topping lift, the latter being a particular nuisance with fully battened mains. With a fully battened main, once set up, the rod kicker needs very little attention.

Another feature of the mainsail, not apparent in the drawings, is that it has a loose foot (with batten) to make removing the sail, complete with battens, a straightforward job.

All control lines and reefing pennants for the mainsail are led aft through clutches to self-tailing winches. To avoid the danger of crew walking on the rolling ropes on deck, the lines are led under a hinged (for access to re-reeve) garage. Halyard and control-line falls are stowed in open-topped coaming lockers. Mainsail reefing is a straightforward two-line remote slab system with luff and leech downhauls.

Deck and cockpit

Although requiring maintenance, on the basis of effectiveness and simplicity, the deck and coachroof are covered with non-skid paint. Treadmaster is used in the cockpit sole and any hazard areas, such as deck hatches, have 3M non-skid strips. The forward hatch opens aft and the saloon hatch forward, offering different combinations for ventilation. Jackstays run from by the cockpit (allowing safety lines to be clipped on before leaving) right up to the foredeck and there are harness eyes in the cockpit.

One slightly innovative feature is the use of vertical rollers on the stemhead cheeks **(Fig 4)** to aid anchor handling when the cable is at an angle. The anchor is designed to stow in the stemhead and is held in place (see **Fig 5**) by means of a chain claw attached to a tensioning bottlescrew. A vertical-

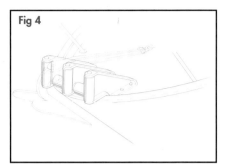

The stemhead fitting incorporates vertical as well as horizontal rollers to allow for easier anchor handling when the cable is at an angle

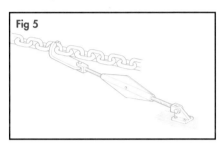

For holding the anchor stowed in the roller a simple tensioning device is used. A chain claw, available from ironmongers or agricultural merchants, is attached via a wide-bodied bottlescrew to a deck eye. Once hooked on, the bottlescrew is tensioned to hold the anchor firmly in position

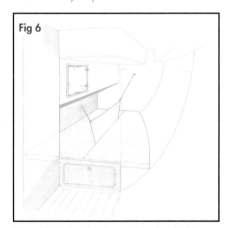

The aftercabin bunk can be divided by a central leecloth to create a sea-berth on either tack. The hatch, which opens into the cockpit well, is large enough to use as an escape hatch

axis windlass and warping drum of the Anchorman type is fitted.

The toerail is a slightly-raised bulwark with enclosed 'Panama' fairleads at the bow. Foredeck cleats are large and well backed with stiffening pads below decks. Deck eyes are fitted between stanchions as strong points and for preventers. There are large dorade ventilators over the galley and heads with bars over them to prevent lines from snagging and for use as grabrails.

The cockpit seats have teak slats, are curved and have double-angled coamings for comfort. The large 'goalpost' at the stern is an integral part of the pushpit and, apart from raising the Decca and Navtex aerials for better reception, it keeps them out of the way of sheets and warps. The stern light, a cockpit light and the Autohelm vane are also mounted here.

The liferaft stows below the after cockpit seat, with a draining gas locker to port and a customised locker for warps, fenders and fuel cans to starboard. The forward cockpit locker is fitted with a sole and a deeply fiddled shelf outboard. It gives access to the oilskin locker which can also be reached from the shower compartment below decks. All cockpit locker lids bed down on neoprene seals. The aftercabin hatch in the cockpit well is larger than usual because it is intended to double as an escape hatch in the event of fire. The primary bilge pump is in the cockpit with a secondary pump in the heads doubling as a shower sump pump.

The companionway 'trough', apart from providing an effective bridgedeck, makes an excellent watchkeeper's seat, and the wooden hatch slides well aft beyond the hatch board to give added shelter below decks.

Accommodation

As in other parts of the boat, we have taken proven ideas and layouts and incorporated them. The three-cabin layout is conventional, all bunks are 6ft 6in or longer and headroom is in excess of 6ft in all cabins. Light laminates and wood (ash) are used throughout to make the interior bright.

The forecabin is intended to be a proper in-harbour cabin, so has standing space and good clothes stowage. The bunks are hinged along their outboard edges, giving easy access to stowage bins for sails and other bulky items below their after ends and sealed lockers with front access ahead of these. The port settee converts to a double and there are lockers both behind and above the backrests.

A fore and aft chart table is fitted, as opposed to the more conventional athwartships type, as it allows for a

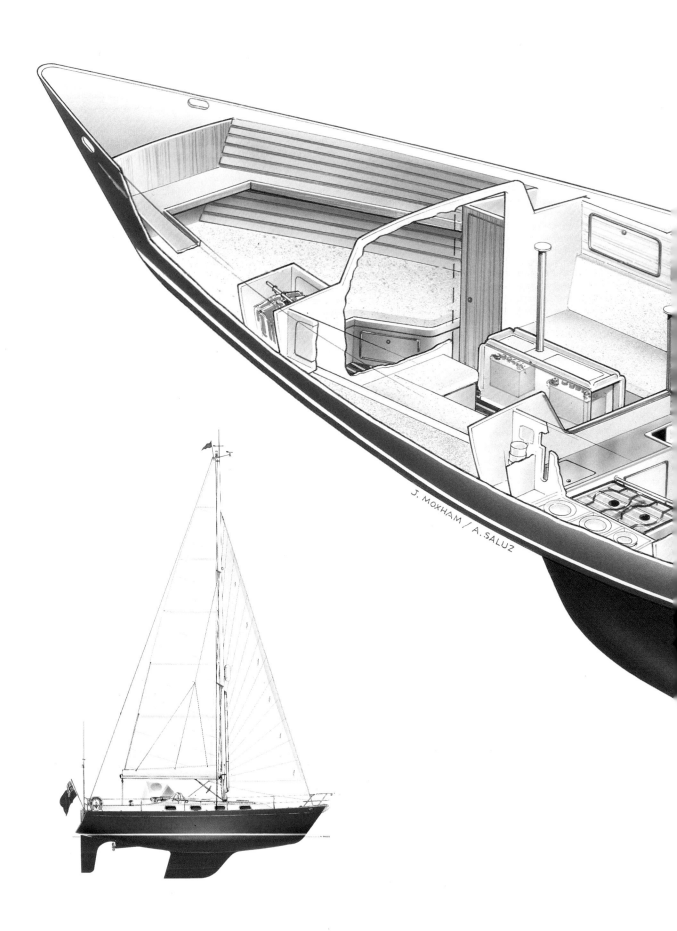

YM OFFSHORE 35

- Forecabin berths hinge along their outboard edges for easy access to large storage bins below
- Service batteries stow in the saloon table for access and to keep their weight amidships
- The U-shaped galley offers security at sea, good work surfaces and stowage
- An instrument and switchboard console runs round three sides of the chart table with a shelf on top for chart tool stowage. Outboard is a large bookshelf and below a storage bin
- The heads has a separate curtained shower cubicle and access to the oilskin locker which can also be reached from the cockpit locker
- The main cockpit locker has a sole and deeply fiddled shelf outboard. The tube contains the dan buoy
- The two quarter cockpit lockers have dedicated stowage for gas bottles (port) and warps, fenders and fuel cans (starboard). The liferaft stows below the after cockpit seat

> Although there is some innovation for the most part, **proven** design and layout features have been incorporated

larger table and on the basis that at sea it is more usual for the navigator to stand. The seat swings out of the way below the table when not in use. An instrument and switchboard console is fitted to three sides of the table with a trough above for navigation tools and there is a large bookshelf outboard of the table.

One of the shortcomings of most galleys is the lack of worktops, so this has been optimised on the YM Offshore 35 by fitting a central peninsula unit for the sinks. This has several advantages. First, the sinks will drain better being near the boat's centreline. Second, the unit offers better security for the cook and, third, it means that two people can work at either side of the sink simultaneously, perhaps one washing, the other drying.

Abaft the cooker, which has two burners, grill and oven with flame failure devices on all, is a large, top-access icebox/fridge and ahead of it a swing top rubbish bin is let into the worktop, making good use of what is otherwise dead space. There is good stowage all round and a cooker crash bar and belt for the cook.

The heads is deliberately large to allow space for a separate, curtained-off shower cubicle. All mouldings are well rounded to allow easy cleaning and prevent the build-up of dirt in crevices. Access to the oilskin locker is at the after end of the cubicle and, if a heater is fitted, an outlet could be directed through this locker for drying.

Like the forecabin, the aftercabin has been designed with a good standing area and plenty of stowage. The bunk is wide and long enough to make a true double and can be divided by means of a central leecloth (see **Fig 6**) to make a good sea-berth. At the foot of this berth are general purpose stowage bins for bedding and the like.

Engine installation

The YM Offshore 35 is fitted with a modern 27hp three-cylinder, fresh/raw water cooled diesel, giving her a hp:displacement ratio of 4.65hp/ton, adequate to maintain hull speed at moderate revs and with enough power for motor-sailing in strong winds.

The 30-gallon stainless fuel tank is fitted under the after double berth, ample for a normal cruise and also sufficient for an extended offshore passage down to Spain or out to the Azores. The fuel system incorporates a two-gallon gravity-feed day tank (for efficiency and fuel monitoring) ahead of twin agglomerator water filters, plumbed so that one can be switched off and drained while the engine is running on passage.

Keen to avoid the potential problems of a P-bracket, the engine was designed to drive a prop-shaft through a conventional shaft log (with Deep-Sea seal). A line-cutter is attached just forward of a fixed two-bladed propeller.

Electrics and instruments

Battery charging is looked at in two ways. Assuming she'll not spend too much time in a marina, shore power has been dropped in deference to simplicity and systems have been designed for self-sufficiency offshore. Conventional alternator charging is supplemented by a TWC regulator whilst the yacht is under way and cruising. A single 18 watt solar panel pampers the batteries in between cruises on the assumption that (in reality) the boat is unlikely to be left with batteries fully charged. A couple of weeks on the mooring will give the solar panel the opportunity to top the batteries right up.

> In the context of the modern production cruising yacht she incorporates all the **features** and **detail** that add up to seaworthiness

The two 100ah domestic batteries are located in the saloon table for two reasons: first, to keep their considerable weight central and, second, to make them accessible for routine maintenance checks. The engine has a dedicated starter 70ah battery in the engine room and charging/service priority is via a heavy-duty, manually operated battery switch.

It would be easy to equip the YM Offshore 35 with a plethora of electronic goodies but, again thinking of maintaining them offshore, power requirements and keeping things as simple as possible, we have kept to fairly comprehensive basics. She is equipped with an Autohelm 2000, Decca and Navtex, all of whose aerials are mounted on the pushpit goalpost. The VHF aerial is mounted at the masthead, although an auxiliary aerial can be fitted on the goalpost in the event of a dismasting.

Wind and water instruments are cockpit-mounted and comprise a speedo/log, echo sounder, wind direction/speed and a Decca repeater.

Conclusion

Like any boat built, the YM Offshore 35 will never be all things to all people. Some readers will disagree with the design and equipment priorities, but it is our view and that of our Advisory Panel that, in the context of the modern production cruising yacht, she incorporates all the features and detail that add up to seaworthiness.

Throughout this book has been an underlying question, 'Are today's yachts any more seaworthy than those of 10 years ago?' The answer is both yes and no.

There were good and bad boats in 1979 and there are good and bad boats today. In terms of materials, structure and the reliability of primary and auxiliary systems, the 1989 yacht is more seaworthy. However, there are some design trends that have led towards a certain type of yacht that is less stable and will be difficult to handle in strong winds. These design characteristics include the trend towards light, beamy, lightly ballasted and shallow hulls. We hope that this book has done something to check the further evolution of design in this direction.

As a final thought, it is worth bearing in mind that the seaworthiness of the yacht itself is only part of the equation. The rest is entirely up to the skills of the skipper and his crew.

INDEX

Anchor 30, 31, 38, 77, 85

Bamar in-mast reefing 27
Bilge 9, 84
Bilge pump 19, 38, 67, 85
Blocks 27
Bow fitting 35, 85
Bridgedeck 31, 84

Chainplates 28
Chart table 37, 39, 40, 88
Cleats 31, 32, 35
Clevis pin 29
Cockpit 8, 19, 21, 22, 30, 31, 32, 82, 84, 85
   locker lids 22, 66, 76, 77, 84
Companionway 8, 19, 20, 40, 52, 66, 67,
.... 76, 84
Cookers 38, 52, 67, 68
Cool box 38
Cruising chute 28
Cutter rig 25

Danbuoy 79, 82
Deck tank fillers 31, 33, 42
Dodgers 32, 33
Dutchman 27

Electrics 22, 38, 39, 52, 67, 68, 70-75 78, 88
   alternator 70, 71, 88
   battery .. 38, 39, 52, 67, 68, 71, 73, 74, 88
   battery switch 71
   connectors 73, 75
   deck plugs 22, 71, 75
   switch panel 71, 73, 74, 88
   TWC regulator 88
   wiring 74, 75, 78
Engine 7, 8, 31, 38, 42-46, 66, 68, 84, 88
   controls 31
   drive belt 43
   P-bracket 43
   propshaft 45, 84, 88

   line cutter 43, 88
   stern gland 43
   water inlet 43
   water pump 43
EPIRB 76
Exhaust pipe 22, 42

Fairlead 35
Fiddles 39
Fire extinguishers 38, 77, 78
Fuel tank 9, 38, 42, 44, 46, 88
Fractional rig 24, 25, 84

Galley 37, 38, 68, 88
Gas system 38, 67, 68, 77, 78, 85
Gaslo meter 77
Guardrails 30, 34

Halyards 29
   wire 29
   rope 29
Handholds 9, 30, 31, 33, 36, 39
   on deck 30, 31, 33
   below 9, 36, 39
Hatches 8, 19, 20, 21, 22, 30, 31, 84, 85
   mainhatch 8, 19, 20, 22, 84
   deck 21, 30, 31, 84, 85
Heads 9, 37, 38, 88
Hull plugs 19, 20, 22, 76

Jackstays 30, 31, 33, 34, 66, 67, 79, 85
Jon Buoy 79

Keel 14, 15, 16, 20, 84,
   design 14, 15, 16, 84
   attachment 20
Kickers 29, 84

Lazyjacks 27, 84
Leecloth 36, 37, 68, 88
Lifebuoy 79

Lifejacket	79
Lifeloop	79
Liferaft	52, 63-65, 80
stowage	35, 65, 85
Lighting	40
Lockerage	31, 37, 38, 39
cockpit	31
below	37, 38, 39
Mainsail	24, 26
fully battened	27, 84
Mainsheet	8
Mast box	35
Non-slip	30, 31, 33, 34, 85
Oilskin locker	88
Outdrives	22
Panic bag	78
Pilot berths	37, 38
Pipe cots	37
Positive flotation	11, 78
Pulpit	30
Radar reflector	79
Reefing	32, 82, 85
Rigging	28, 29
standing	28, 29
running	29
Rigging lubricants	29
Rigging screws	28
Rigging terminals	28
Roller reefing	24, 25, 26, 28, 29, 82, 84
headsail	24, 25, 26, 29, 82, 84
in-mast	28
Rudder	14, 16, 84
Safety harness	66, 79
anchorage	30, 33, 60-62, 66, 67, 79, 85
harness clips	33, 60
Saloon table	39
Seaberth	36, 37, 39, 85, 88
Sea Sure Fastnet Latch	20
Sheet car	26
Shroud terminals	28, 29
Sinks	38, 88
Skin fittings	22, 44, 67, 84
Solar panel	88
Sole	40, 52
Spinnaker	28
Split pins	29, 80
Sprayhood	32
Stack system	84
Stanchions	30, 31
Steering	32, 52
Stern gland	22
Storm jib stay	26
Subrella	77
Tinker inflatable	80
Toerail	34
Toggles	28
Towing points	30
Treadmaster	40
Unsink	78
Upholstery	37
Ventilation	22, 36, 40, 82, 84, 85
VHF	77, 80
Washboards	19, 20, 22, 52, 84
Water ballast	13
Water tank	38, 85
Watertight bulkhead	11, 19, 79
Winches	8, 14, 31
Windlass	35
Windows and ports	8, 20, 21, 52, 84
Window storm boards	19, 21
Yacht Tidy	32

Other books published by Yachting Monthly

West Country Cruising
£13.95
- Contents fully revised and updated
- Over 150 additional photographs
- New section, coastal views and harbour approaches

In just over a year, the first edition of Mark Fishwick's pilot to the coasts, harbours and rivers of the West Country from the River Exe to Lands End and the north Cornish harbours sold out. In that period it became established as the definitive West Country Pilot book and no other publication comes close to it in terms of detail, the lavish use of colour photographs and aerial views and its wealth of local background. The **new edition** has been fully updated, contains a new section with over 150 photographs of harbour and coastal views and approaches.

East Coast Rivers
£9.95
- new Navaid Review edition including all the latest changes to the River Crouch buoyage

East Coast Rivers is Yachting Monthly's longest established pilot book, written by East Coast veteran Jack Coote. Covering the rivers, harbours and swatchways from Southwold to the Swale, East Coast Rivers is the most comprehensive and up-to-date pilot book of this area.

East Coast Rivers from the Air
£4.95

Companion to East Coast Rivers and also written by Jack Coote, this handy book contains excellent aerial views of the East Coast as an invaluable aid to navigation.

Classic Passages
£9.95

As a guide to cruise and passage planning in the English Channel, Normandy, North and South Brittany, the Channel Islands and the Isles of Scilly, Classic Passages has no equal. Published in Association with the Royal Cruising Club's Pilotage Foundation the book sets a new level in cruising information.

Sailpower
£2.50

Why sail badly? Sailpower, written by expert Bunty King, tells you how to make the most of a cruising boat's rig and sails. 28 pages packed with practical advice

The Best of Looking Around
£1.95

What pub opened fire on a Royal Naval Ship? Who said after his first circumnavigation and before two more 'I'll never do it again'? The answers and many more snippets culled from the columns of Looking Around during the period Bill Beavis was its author are in this pocket sized book, the perfect stocking filler.